SOLVING AMERICA'S PROBLEMS: A CALL FOR UNITY AND ACTION

I0427232

PART 1:

ADDRESSING THE WEALTH GAP: SOLUTIONS FOR CREATING A MORE EQUITABLE ECONOMY

BY
HENRY E. PARKINS

1

COPYRIGHT PAGE

TABLE OF CONTENTS

INTRODUCTION

SOLVING AMERICA'S PROBLEMS: A CALL FOR UNITY AND ACTION

In the landscape of contemporary America, a myriad of challenges threatens the very fabric of our society. From the staggering wealth gap to the soaring costs of healthcare, the scarcity of affordable housing to the looming specter of climate change, our nation stands at a critical crossroads. The issues we face are not merely isolated concerns; they are interconnected threads woven into the complex tapestry of American life.

At the heart of our nation's strife lies the widening chasm of economic inequality. The growing wealth gap has become a chasm so vast that it threatens the very essence of the American dream. While a privileged few amass unprecedented fortunes, countless others struggle to make ends meet, trapped in cycles of poverty and disenfranchisement. Part I of our discourse will delve into this pressing issue, exploring solutions for creating a

8

more equitable economy that uplifts all Americans, not just a select few.

Compounding the challenges of economic disparity is the relentless rise in healthcare costs, a burden that weighs heavily on the shoulders of families across the nation. As medical expenses spiral out of control, access to essential healthcare services becomes a luxury rather than a fundamental right. In Part II, we will navigate the labyrinth of healthcare reform, charting a course toward improved access and affordability for all Americans, regardless of socioeconomic status.

Meanwhile, the dream of secure and affordable housing remains elusive for far too many Americans. As rents skyrocket and homeownership slips further out of reach, families find themselves teetering on the brink of homelessness, trapped in a vicious cycle of housing insecurity. Part III of our discussion will explore solutions aimed at making housing more accessible and affordable, ensuring that every American has a place to call home.

And looming on the horizon, casting a shadow of uncertainty over future generations, is the existential threat of climate change. As temperatures rise and natural disasters proliferate, the urgency of addressing this global crisis has never been clearer. In Part IV, we will confront the daunting challenge of climate change head-on, exploring innovative solutions for safeguarding our planet and securing a sustainable future for all.

Yet, amidst the cacophony of discord and despair, there remains a glimmer of hope a beacon of unity and collective action. Now more than ever, we must set aside our differences and stand together in pursuit of a brighter tomorrow. In the conclusion of our discourse, we issue a fervent call to arms, rallying Americans from all walks of life to join forces in the fight for a better, more equitable future. For in unity lies our strength, and in action lies our salvation. Together, let us rise to meet the challenges that confront us, forging a path toward a more just and prosperous America for generations to come.

Part I: Addressing the Wealth Gap - Solutions for Creating a More Equitable Economy

The growing wealth gap in America stands as a stark reminder of the systemic inequalities that plague our society. As the chasm widens between the affluent and the impoverished, the promise of upward mobility dims for millions of hardworking Americans. This disparity not only undermines the principles of fairness and justice upon which our nation was founded but also erodes the very foundation of our democracy.

Acknowledging Structural Inequities:

Central to understanding the wealth gap is the recognition of structural inequities embedded within our economic system. While some amass unimaginable fortunes through a combination of privilege and opportunity, others are relegated to the margins, their potential stifled by systemic barriers that perpetuate cycles of poverty and exclusion.

Income Disparities and Wage Stagnation:

A key driver of the wealth gap is the persistent disparity in income distribution and the stagnation of wages for the working class. Despite increases in worker productivity and corporate profits, wages have failed to keep pace, widening the chasm between those at the top and those struggling to make ends meet.

Racial and Gender Disparities:

The wealth gap is further exacerbated by racial and gender disparities that disproportionately impact marginalized communities. Historical injustices, discriminatory practices, and institutionalized racism have entrenched inequalities, perpetuating a cycle of economic disadvantage that spans generations.

Corporate Power and Influence: The concentration of wealth and power in the hands of a few corporate behemoths exacerbates the wealth gap, consolidating economic control and stifling competition. As corporate interests dictate policy agendas and shape regulatory frameworks, the voices of ordinary Americans are drowned out, further entrenching disparities in wealth and opportunity.

Tax Reform and Redistribution:

Meaningful reform of the tax system is essential to addressing the wealth gap, ensuring that the burden of taxation is equitably distributed and that the wealthiest individuals and corporations pay their fair share. By implementing progressive tax policies and closing loopholes that benefit the ultra-rich, we can generate the resources needed to invest in education, infrastructure, and social programs that uplift the most vulnerable members of society.

Investing in Human Capital: To bridge the wealth gap, we must invest in human capital, providing equitable access to education, job training, and healthcare. By expanding opportunities for upward mobility and leveling the playing field, we can empower individuals to reach their full potential and contribute meaningfully to our economy and society.

Promoting Economic Justice and Equity: At its core, addressing the wealth gap requires a commitment to economic justice and equity. By fostering an inclusive economy that values the dignity and worth of every individual, we can create a society where prosperity is shared by all, not hoarded by the few.

As we confront the challenge of the growing wealth gap, let us heed the call for unity and action, working together to dismantle the barriers that divide us and forge a path toward a more just and equitable future for all Americans.

Part II: Healthcare Reform - Solutions for Improving Access and Affordability

The relentless rise in healthcare costs represents a formidable barrier to the well-being of millions of Americans, undermining the promise of affordable and accessible healthcare for all. As medical expenses spiral out of control, families are forced to grapple with impossible choices, sacrificing their financial security in the pursuit of essential medical care. To confront this pressing issue, we must chart a course toward comprehensive healthcare reform that prioritizes the needs of patients over the profit margins of corporate interests.

Escalating Costs of Medical Services:

At the heart of the healthcare cost crisis lies the escalating expenses associated with medical services, from routine check-ups to life-saving treatments. As the price of prescription drugs skyrockets and hospital bills mount, many Americans find themselves burdened with insurmountable debt, forced to choose between necessary healthcare and other basic necessities.

Inadequate Insurance Coverage:

Despite advances in healthcare coverage through programs like Medicare and Medicaid, millions of Americans remain uninsured or underinsured, leaving them vulnerable to financial hardship in the event of illness or injury. High deductibles, co-payments, and out-of-pocket expenses further compound the financial strain faced by individuals and families struggling to access essential healthcare services.

Pharmaceutical Industry Profiteering: The exorbitant prices

charged by pharmaceutical companies for life-saving medications represent a significant driver of rising healthcare costs. From insulin to cancer treatments, the cost of prescription drugs continues to soar, placing an intolerable burden on patients and straining the resources of healthcare systems nationwide.

Administrative Waste and Bureaucratic Overhead: The

complex web of administrative waste and bureaucratic overhead inherent in our healthcare system contributes to inefficiencies that drive up costs without improving patient outcomes. Streamlining administrative processes and reducing bureaucratic red tape can help redirect resources toward patient care, ensuring that every dollar is spent wisely and efficiently.

Addressing Social Determinants of Health:
Recognizing that health outcomes are influenced by a myriad of social, economic, and environmental factors, we must adopt a holistic approach to healthcare reform that addresses the underlying determinants of health. Investing in preventive care, promoting healthy lifestyles, and addressing social disparities can help mitigate the need for costly medical interventions down the line.

Expanding Access to Affordable Care:
Expanding access to affordable healthcare must be a cornerstone of any comprehensive reform effort. By strengthening and expanding existing healthcare programs, implementing universal coverage options, and enhancing consumer protections, we can ensure that every American has access to the care

16

they need, when they need it, without fear of financial ruin.

Fostering Innovation and Collaboration:

Embracing innovation and fostering collaboration among healthcare providers, insurers, policymakers, and community stakeholders is essential to driving meaningful change in our healthcare system. By leveraging technology, promoting evidence-based practices, and prioritizing patient-centered care, we can create a more efficient, responsive, and equitable healthcare system for all Americans.

As we confront the challenge of rising healthcare costs, let us unite in our commitment to ensuring that healthcare remains a fundamental human right, not a privilege reserved for the wealthy and well-connected. By working together to enact meaningful reforms that prioritize affordability, accessibility, and quality of care, we can build a healthier, more prosperous future for generations to come.

Part III: Housing Affordability Solutions for Making Housing More Accessible and Affordable

The lack of affordable housing stands as a profound challenge to the fabric of American society, casting a shadow of uncertainty over the lives of millions of individuals and families. As rents soar and homeownership slips further out of reach, the dream of secure and stable housing becomes increasingly elusive, pushing vulnerable populations to the margins of society. To address this pressing issue, we must embark on a concerted effort to make housing more accessible and affordable for all Americans.

Rising Housing Costs: Across the nation, housing costs continue to escalate, outpacing wage growth and placing immense strain on household budgets. As demand for housing outstrips supply in many urban centers, rents skyrocket, pricing out low- and middle-income families and exacerbating homelessness and housing insecurity.

18

Shortage of Affordable Housing Units: The shortage of affordable housing units represents a critical bottleneck in efforts to address the housing affordability crisis. Limited availability of subsidized housing options and a dearth of affordable rental properties leave millions of Americans struggling to find safe, decent, and affordable housing options, forcing many into overcrowded or substandard living conditions.

Gentrification and Displacement: Gentrification and displacement further compound the challenges of housing affordability, as historically marginalized communities are pushed out of their neighborhoods by rising property values and redevelopment efforts. As vibrant communities are reshaped by waves of gentrification, longtime residents are left without affordable housing options and face the prospect of displacement from the places they call home.

Barriers to Homeownership: For many Americans, the dream of homeownership remains just that a dream due to a host of barriers, including

19

stagnant wages, high down payment requirements, and limited access to affordable mortgage financing. As a result, homeownership rates among low- and moderate-income households continue to decline, exacerbating wealth disparities and perpetuating cycles of poverty and inequality.

Housing Discrimination and Segregation:
Persistent housing discrimination and segregation perpetuate disparities in access to affordable housing, particularly for communities of color and other marginalized groups. Discriminatory lending practices, exclusionary zoning policies, and systemic barriers to fair housing perpetuate segregation and limit housing options for those most in need.

Investing in Affordable Housing Solutions:
Addressing the lack of affordable housing requires a multifaceted approach that includes robust investments in affordable housing development, preservation, and rehabilitation. By leveraging public-private partnerships, expanding tax incentives, and allocating resources to support affordable housing initiatives, we can increase the supply of

20

affordable housing units and create pathways to homeownership for aspiring homeowners.

Promoting Equitable Housing Policies:

Promoting equitable housing policies is essential to ensuring that all Americans have access to safe, decent, and affordable housing options. By dismantling discriminatory practices, promoting inclusive zoning policies, and combating housing segregation, we can create more inclusive and vibrant communities where everyone has the opportunity to thrive.

As we confront the challenge of housing affordability, let us unite in our commitment to building a more equitable and inclusive society where every individual and family has access to safe, stable, and affordable housing. By working together to implement bold and innovative solutions, we can transform the dream of homeownership into a reality for millions of Americans and build stronger, more resilient communities for generations to come.

Part IV: Climate Change - Solutions for Addressing the Climate Crisis

The specter of climate change looms large, casting a shadow of uncertainty over the future of our planet and the well-being of generations to come. As temperatures rise, sea levels swell, and extreme weather events become increasingly frequent and severe, the urgency of addressing the climate crisis has never been more apparent. To confront this existential threat, we must unite in our resolve to implement bold and transformative solutions that safeguard our planet and secure a sustainable future for all.

Rising Global Temperatures: The steady rise in global temperatures is one of the most visible and alarming manifestations of climate change. As greenhouse gas emissions from human activities continue to accumulate in the atmosphere, the Earth's climate system is pushed beyond its limits, leading to a cascade of ecological disruptions and environmental catastrophes.

Extreme Weather Events: The impacts of climate change are felt acutely through the increasing frequency and intensity of extreme weather events, including hurricanes, wildfires, droughts, and floods. These events wreak havoc on communities, causing loss of life, destruction of property, and disruption of vital infrastructure, exacerbating social and economic disparities and straining emergency response systems to their breaking point.

Melting Polar Ice Caps and Rising Sea Levels: The melting of polar ice caps and the consequent rise in sea levels pose grave threats to coastal communities and low-lying regions around the world. As sea levels continue to rise, vulnerable coastal ecosystems are inundated, freshwater supplies are contaminated, and millions of people are displaced from their homes, exacerbating global migration crises and fueling social and political instability.

Loss of Biodiversity and Ecosystem Degradation: Climate change is driving the loss of biodiversity and the degradation of vital ecosystems at an unprecedented rate. From coral reefs to rainforests, ecosystems that sustain life on Earth are under siege, threatening the stability of entire ecosystems and jeopardizing the survival of countless plant and animal species.

Impact on Public Health: Climate change poses significant risks to public health, exacerbating respiratory illnesses, heat-related illnesses, and the spread of infectious diseases. Vulnerable populations, including children, the elderly, and low-income communities, are disproportionately affected by climate-related health impacts, exacerbating existing health disparities and straining healthcare systems already stretched thin by the demands of the pandemic.

Transition to Clean Energy:

Transitioning to a clean energy economy is essential to mitigating the impacts of climate change and reducing our dependence on fossil fuels. By investing in renewable energy sources such as solar, wind, and hydroelectric power, we can decrease our carbon footprint, create green jobs, and foster economic growth while protecting the health and well-being of future generations.

Adaptation and Resilience:

Building resilience to the impacts of climate change is paramount to protecting communities and ecosystems from the inevitable challenges that lie ahead. By investing in climate-resilient infrastructure, implementing nature-based solutions, and adopting sustainable land and water management practices, we can enhance our ability to adapt to changing environmental conditions and minimize the risks posed by climate-related disasters.

International Cooperation and Leadership:

Addressing the global challenge of climate change requires international cooperation and leadership on a scale never before seen. By recommitting to the Paris Agreement, strengthening multilateral partnerships, and mobilizing resources to support climate action in developing countries, we can forge a path toward a more sustainable and equitable future for all.

As we confront the existential threat of climate change, let us heed the call for unity and action, working together to

25

protect our planet and preserve the wonders of the natural world for future generations. By embracing innovation, fostering collaboration, and embracing our shared responsibility to safeguard the Earth, we can rise to the challenge of the climate crisis and build a brighter, more sustainable future for all.

PART 1:

ADDRESSING THE WEALTH GAP: SOLUTIONS FOR CREATING A MORE EQUITABLE ECONOMY

INTRODUCTION

In the landscape of American society, the chasm between the affluent and the impoverished has grown into a chasm of staggering proportions. This chasm, known as the wealth gap, represents not just a statistical disparity, but a stark manifestation of systemic inequities entrenched within the fabric of the nation's economic structure. As wealth becomes increasingly concentrated in the hands of a few, millions of Americans struggle to make ends meet, perpetuating cycles of poverty and stifling social mobility.

"Addressing the Wealth Gap: Solutions for Creating a More Equitable Economy in the US" delves into the heart of this pressing issue, dissecting its roots, unraveling its complexities, and presenting a roadmap toward a more just and inclusive economic future. Within these pages, we embark on a journey to understand the multifaceted nature of the wealth gap and explore viable strategies for bridging the divide.

The urgency of this endeavor cannot be overstated. The repercussions of widening economic disparities reverberate through every aspect of society, eroding the very foundation of democracy and social cohesion. Yet, amidst the bleakness of inequality, there exists a glimmer of hope—a collective recognition of the imperative to confront this challenge head-on and forge a path toward a fairer, more prosperous nation for all.

This book is not merely an academic exercise or a catalog of grievances; it is a call to action a rallying cry for policymakers, activists, businesses, and citizens alike to join forces in the pursuit of economic justice. By examining the structural barriers that perpetuate the wealth gap and exploring innovative solutions, we aim to empower readers with knowledge, inspire meaningful dialogue, and catalyze transformative change.

Throughout these pages, we will confront uncomfortable truths, challenge entrenched ideologies, and envision a future where opportunity is not a privilege reserved for the few, but a fundamental right afforded to all. From reimagining fiscal policies to dismantling systemic racism, from investing in education to fostering inclusive entrepreneurship, we will explore a spectrum of strategies aimed at reshaping the economic landscape and fostering a more equitable society.

The journey ahead will not be easy, nor will it be without setbacks and obstacles. Yet, in the face of adversity, we find the resilience of the human spirit and the power of collective action. Together, we have the capacity to transcend divisions, defy inertia, and build a future where every individual can thrive, irrespective of their background or circumstance.

As we embark on this voyage, let us heed the words of Martin Luther King Jr., who famously proclaimed, "Injustice anywhere is a threat to justice everywhere." Let us rise to the challenge of our times with courage, compassion, and unwavering resolve. For in the struggle for economic equity lies the promise of a brighter tomorrow—for our nation, for our communities, and for generations yet to come.

Welcome to "Addressing the Wealth Gap: Solutions for Creating a More Equitable Economy in the US." Together, let us embark on a journey of discovery, empowerment, and transformation.

Definition and Significance of the Wealth Gap

The wealth gap, also known as income inequality or economic disparity, refers to the unequal distribution of wealth and assets among individuals or groups within a society. It encompasses disparities in income, assets, savings, investments, and access to opportunities, resulting in wide variations in economic well-being and living standards.

At its core, the wealth gap represents a fundamental imbalance in the distribution of resources and opportunities, wherein a disproportionate share of wealth and prosperity accrues to a select few while leaving a significant portion of the population marginalized and economically disenfranchised. This disparity is often manifested in stark contrasts between the affluent elite and marginalized communities, exacerbating social divisions and perpetuating cycles of poverty and exclusion.

The significance of the wealth gap extends far beyond mere economic statistics; it constitutes a profound social, political, and moral challenge with profound implications for the fabric of society. At its most basic level, the wealth gap undermines the principles of fairness, justice, and equality that lie at the heart of democratic ideals. It erodes social cohesion, fosters resentment and distrust, and undermines the social contract that binds communities together.

Moreover, the wealth gap poses a formidable barrier to upward mobility and economic opportunity, effectively trapping millions of individuals and families in cycles of poverty and deprivation. It limits access to quality education, healthcare, housing, and other essential services, perpetuating intergenerational cycles of disadvantage and depriving entire communities of the chance to fulfill their potential.

From a macroeconomic perspective, the wealth gap undermines the stability and sustainability of the economy, hindering long-term growth and prosperity. Concentrated wealth at the top limits consumer spending, dampens demand for goods and services, and stifles innovation and entrepreneurship, thereby impeding economic dynamism and vitality.

Furthermore, the wealth gap exacerbates social tensions and undermines the fabric of democracy, fueling resentment, polarization, and social unrest. It breeds disillusionment with political institutions and undermines faith in the fairness and legitimacy of the economic system, creating fertile ground for social upheaval and political instability.

In light of these profound implications, addressing the wealth gap is not merely a matter of economic policy, but a moral imperative and a fundamental challenge to the values of equality, justice, and opportunity upon which our society is built. It demands bold and comprehensive solutions that address the root causes of inequality, dismantle systemic barriers to economic mobility, and foster a more inclusive and equitable economy for all.

In the pages that follow, we will explore the multifaceted nature of the wealth gap, examine its underlying causes and consequences, and propose actionable strategies for building a more just, prosperous, and equitable society. By confronting the wealth gap head-on and working collectively to address its root causes, we can forge a path toward a future where every individual has the

33

opportunity to thrive and contribute to the common good.

Historical Context and Evolution of Economic Inequality in the US

The story of economic inequality in the United States is deeply intertwined with the nation's history, shaped by a complex tapestry of social, political, and economic forces that have evolved over centuries. From the earliest days of colonization to the present day, the trajectory of economic inequality has been marked by periods of expansion and contraction, progress and regression, but its enduring legacy remains deeply entrenched in the fabric of American society.

Colonial Era and Agrarian Economy:

Economic inequality in the US can be traced back to the colonial era, where wealth and power were concentrated in the hands of a landed elite, often composed of colonial aristocrats and plantation owners. The agrarian economy of the South, built upon the institution of slavery, perpetuated stark disparities in wealth and opportunity, with enslaved Africans and their descendants relegated to the margins of society.

Industrial Revolution and Gilded Age:

The 19th century witnessed the dawn of the Industrial Revolution, transforming the American economy from agrarian to industrial.

While the period saw unprecedented economic growth and technological innovation, it also gave rise to rampant inequality, epitomized by the era of the Gilded Age.

Robber barons amassed vast fortunes through monopolistic practices, while workers toiled in deplorable conditions for meager wages, giving rise to widespread social unrest and labor activism.

Progressive Era and New Deal Reforms:

The early 20th century saw the emergence of the Progressive Movement, which sought to address the social and economic inequities of the Gilded Age.

Progressive reforms, including antitrust legislation, labor protections, and the introduction of the progressive income tax, aimed to curb the excesses of corporate power and promote greater economic fairness.

The New Deal programs of the 1930s, enacted in response to the Great Depression, ushered in a new era of social welfare policies aimed at alleviating poverty, stimulating economic recovery, and fostering greater economic security for all Americans.

Post-War Prosperity and the Rise of the Middle Class:

The post-World War II era witnessed unprecedented economic growth and prosperity, characterized by rising wages, expanding opportunities, and the emergence of a burgeoning middle class.

Government investments in infrastructure, education, and healthcare, coupled with robust labor protections and social welfare programs, contributed to a more equitable distribution of wealth and opportunity.

Neoliberal Era and Rising Inequality:

The latter half of the 20th century saw the rise of neoliberal economic policies, characterized by deregulation, privatization, and the erosion of social safety nets.

The Reagan Revolution and subsequent administrations prioritized market-driven solutions and tax cuts for the wealthy, leading to a widening wealth gap and declining economic mobility for low- and middle-income Americans.

Globalization, technological advancements, and the decline of organized labor further exacerbated disparities in income and wealth, consolidating economic power in the hands of a wealthy elite.

Contemporary Challenges and Persistent Inequities:

In the 21st century, economic inequality in the US has reached unprecedented levels, with the wealthiest 1% capturing a disproportionate share of income and wealth.

Racial and gender disparities persist, exacerbating the wealth gap and perpetuating systemic injustices that disproportionately affect communities of color and marginalized groups.

The COVID-19 pandemic has laid bare the deep-seated inequities within the US economy, exposing vulnerabilities and exacerbating disparities in access to healthcare, employment, and economic opportunity.

As we navigate the complexities of economic inequality in the US, it is imperative to understand its historical roots and evolution, recognizing the enduring legacy of systemic injustices that continue to shape the contours of our society. By confronting the historical context of economic inequality and acknowledging the structural barriers that perpetuate it, we can chart a path toward a more just, equitable, and inclusive economy for all Americans.

Purpose and Scope of the Book

Addressing the Wealth Gap:

Solutions for Creating a More Equitable Economy in the US" is a comprehensive examination of one of the most pressing challenges facing American society today the pervasive and widening wealth gap. Rooted in a deep commitment to social justice and economic fairness, this book seeks to illuminate the complexities of economic inequality in the United States and explore actionable strategies for building a more just and inclusive economy.

The primary purpose of this book is threefold:

To Provide Understanding: The book endeavors to provide readers with a nuanced understanding of the wealth gap, its historical antecedents, and its multifaceted manifestations in contemporary American society. By examining the root causes and systemic barriers that perpetuate economic inequality, readers will gain insight into the structural forces shaping the economic landscape and the lived experiences of millions of Americans.

To Explore Solutions: **Central to the book's mission is the exploration of viable solutions and innovative strategies for addressing the wealth gap. Drawing upon a diverse array of perspectives, research findings, and real-world examples, the book offers a roadmap for policymakers, activists, businesses, and citizens to collaborate in the pursuit of economic justice. From policy reforms to community-driven initiatives, from corporate accountability to grassroots advocacy, the book presents a spectrum of approaches**

aimed at fostering greater equity and opportunity for all.

To Inspire Action: Ultimately, "Addressing the Wealth Gap" aspires to inspire action and catalyze transformative change. By amplifying voices of advocacy, highlighting success stories, and empowering readers with knowledge and agency, the book seeks to mobilize individuals and communities in the collective endeavor to build a more equitable and inclusive economy. Through informed dialogue, strategic collaboration, and sustained advocacy, we can confront the wealth gap head-on and work towards a future where every individual has the opportunity to thrive and fulfill their potential.

The scope of the book is broad yet focused, encompassing a comprehensive analysis of the wealth gap from historical, economic, social, and political perspectives. It delves into the root causes and consequences of economic inequality, explores the intersectionality of race, gender, and class, and examines the role of various institutions and stakeholders in perpetuating or mitigating disparities.

Furthermore, the book offers a forward-looking perspective, examining emerging trends, challenges, and opportunities in the quest for economic justice. It underscores the imperative of collective action and underscores the urgency of addressing the wealth gap as a moral imperative, an economic imperative, and a fundamental challenge to the values of democracy and equality.

In summary, "Addressing the Wealth Gap" is a call to action a testament to our shared commitment to building a more just, equitable, and inclusive society. It invites readers to join in the collective endeavor to confront economic inequality, dismantle systemic barriers, and forge a path toward a future where opportunity is not a privilege but a fundamental right for all Americans.

CHAPTER 1

Understanding the Wealth Gap

The wealth gap, a pervasive and deeply entrenched phenomenon within American society, defies simple explanation and transcends mere economic statistics. At its core, the wealth gap represents a stark manifestation of systemic inequalities that have shaped the economic landscape of the United States for generations. To truly comprehend the complexities of the wealth gap, it is essential to explore its multifaceted nature and the myriad factors that contribute to its existence.

Factors Contributing to the Wealth Gap

a. Income Inequality: One of the primary drivers of the wealth gap is income inequality, whereby individuals and households earn vastly disparate incomes based on factors such as education, occupation, and systemic biases within the labor market.

42

b. Racial Disparities: Race continues to be a significant determinant of wealth and economic opportunity in the United States. Historic injustices such as slavery, segregation, and discriminatory policies have resulted in persistent disparities in wealth accumulation between white Americans and communities of color.

c. Educational Disparities: Access to quality education is a critical determinant of economic success and wealth accumulation. However, systemic inequalities in educational funding, resources, and opportunities perpetuate disparities in academic achievement and economic mobility.

d. Systemic Barriers: Structural inequities embedded within the fabric of society, including discriminatory lending practices, unequal access to healthcare, and disparities in criminal justice, create barriers to economic advancement and perpetuate cycles of poverty and inequality.

Impact of the Wealth Gap

a. Social Cohesion: The widening wealth gap undermines social cohesion and erodes trust in institutions, fostering

43

divisions along socioeconomic lines and exacerbating social tensions within communities.

b. Economic Growth: Economic inequality impedes long-term economic growth and prosperity by limiting consumer spending, dampening demand for goods and services, and hindering investment in human capital and innovation.

c. Political Stability: Inequities in wealth and political power undermine the democratic principles of representation and equality, fueling disillusionment with the political system and eroding faith in democratic institutions.

Case Studies and Data Analysis: Through empirical research and case studies, it is possible to glean insights into the dynamics of the wealth gap and its consequences for individuals, families, and society at large. By examining trends in wealth distribution, income mobility, and intergenerational wealth transfer, researchers can better understand the root causes and implications of economic inequality.

In summary, understanding the wealth gap requires a holistic analysis of its underlying drivers, its impact on society, and the structural barriers that perpetuate its existence. By confronting the complexities of economic inequality and fostering dialogue around solutions, we can begin to address the root causes of the wealth gap and work towards building a more equitable and inclusive economy for all Americans.

Factors Contributing to the Wealth Gap

The wealth gap in the United States is the result of a complex interplay of economic, social, and historical factors that have shaped the distribution of wealth and opportunity within society. Understanding these contributing factors is essential for devising effective strategies to address the wealth gap and promote greater economic equity. Here are some key factors:

Income Inequality

Income inequality is a fundamental driver of the wealth gap. Disparities in wages and salaries across different sectors and occupations contribute to unequal wealth accumulation over time. Factors such as globalization, technological change, and the decline of organized labor have exacerbated income inequality, with higher

45

earners capturing a disproportionate share of economic gains.

Racial and Ethnic Disparities

Racial and ethnic disparities play a significant role in perpetuating the wealth gap. Historical injustices such as slavery, segregation, and discriminatory policies have led to persistent disparities in wealth accumulation between white Americans and communities of color. Structural racism in areas such as housing, education, employment, and criminal justice further exacerbates these disparities and limits economic mobility for marginalized groups.

Educational Disparities

Access to quality education is a critical determinant of economic opportunity and wealth accumulation. However, disparities in educational resources, funding, and opportunities perpetuate inequalities in academic achievement and limit economic mobility, particularly for low-income and minority students. Unequal access to higher education further exacerbates the wealth gap by limiting opportunities for upward social mobility.

Inter-Generational Wealth Transfer

Intergenerational wealth transfer plays a significant role in perpetuating wealth inequality. Inheritances, gifts, and family wealth transfers disproportionately benefit affluent families, perpetuating dynastic wealth and exacerbating disparities between affluent and low-income households. Limited access to intergenerational wealth transfer mechanisms further widens the wealth gap and reinforces existing inequalities.

Systemic Barriers and Discrimination

Structural barriers and systemic discrimination within institutions and systems contribute to the perpetuation of the wealth gap. Discriminatory lending practices, unequal access to credit and financial services, disparities in healthcare access and outcomes, and biases within the criminal justice system disproportionately affect marginalized communities and limit their ability to accumulate wealth.

Tax Policies and Economic Structures

Tax policies and economic structures can either exacerbate or mitigate wealth inequality. Regressive tax policies, loopholes, and preferential treatment for high-income individuals and corporations contribute to widening wealth disparities by allowing the wealthiest individuals and corporations to accumulate wealth at the expense of working families and low-income individuals. Inequitable economic structures, such as the concentration of wealth in the financial sector and speculative activities, further exacerbate wealth inequality.

Understanding these factors is essential for developing comprehensive strategies to address the wealth gap and promote economic equity. By addressing root causes such as income inequality, racial disparities, educational inequities, intergenerational wealth transfer, systemic barriers, and inequitable tax policies, policymakers, businesses, and communities can work together to create a more equitable economy that benefits all members of society.

Income Inequality

Income inequality stands as one of the most significant drivers of the wealth gap in the United States, reflecting disparities in earnings and wages among individuals and households across the socioeconomic spectrum. The issue is multifaceted, rooted in a complex interplay of economic, social, and policy factors that shape the distribution of income within society. Understanding the dynamics of income inequality is crucial for devising effective solutions to address the wealth gap and foster a more equitable economy.

Economic Forces: Income inequality is influenced by a variety of economic forces, including technological change, globalization, and shifts in labor market dynamics. Technological advancements and automation have transformed industries and reshaped the nature of work, leading to the polarization of employment opportunities and exacerbating wage differentials between high-skilled and low-skilled workers. Globalization has contributed to the outsourcing of jobs and the erosion of wages in certain sectors, further widening income disparities.

Wage Stagnation: Despite overall economic growth, wages for many workers have stagnated or grown at a slower pace compared to productivity gains and corporate profits. This phenomenon, often referred to as wage stagnation, has contributed to the widening gap between the earnings of workers and the incomes of corporate executives and shareholders. Structural factors such as declining unionization rates, weakened labor protections, and the erosion of the minimum wage have undermined workers' bargaining power and contributed to stagnant wages for many.

Executive Compensation: The exponential growth of executive compensation in recent decades has played a significant role in exacerbating income inequality. The compensation packages of CEOs and top executives at large corporations have soared to unprecedented levels, far outpacing the wage growth of average workers. This trend reflects not only changes in corporate governance structures but also broader shifts in societal norms and values

50

regarding income distribution and executive pay.

Tax Policies and Regulatory Frameworks:
Tax policies and regulatory frameworks can either mitigate or exacerbate income inequality. Historically, tax policies have played a role in redistributing income and wealth through progressive taxation and social welfare programs. However, changes in tax laws and regulatory frameworks over the past several decades have favored the wealthy and corporations, contributing to the concentration of income and wealth among the top income earners. Tax cuts for high-income individuals, capital gains, and corporate profits have disproportionately benefited the wealthiest segments of society, further widening income disparities.

Educational Attainment and Skills:
Education remains a critical determinant of income and earning potential in the modern economy. Individuals with higher levels of educational attainment and specialized skills often command higher wages and have greater access to high-paying jobs.

However, disparities in educational opportunities and access to quality education perpetuate income inequalities, particularly for marginalized communities and low-income households. Addressing disparities in educational attainment and investing in workforce development and lifelong learning initiatives are essential for reducing income inequality and promoting economic mobility.

Addressing income inequality requires a comprehensive approach that encompasses policy interventions, structural reforms, and investments in education, workforce development, and social welfare programs. By addressing the root causes of income inequality and fostering greater economic opportunity for all Americans, policymakers, businesses, and civil society can work together to build a more equitable economy that benefits everyone.

Racial Disparities

Racial disparities constitute a critical dimension of the wealth gap in the United States, reflecting deep-seated inequities rooted in the nation's history of slavery, segregation, and systemic discrimination.

52

Despite progress toward racial equality, disparities in wealth, income, education, employment, and access to opportunities persist, perpetuating structural barriers that hinder the economic advancement of communities of color. Understanding the complexities of racial disparities is essential for devising effective strategies to address the wealth gap and promote racial and economic justice.

Historical Legacy: The legacy of slavery, Jim Crow laws, redlining, and other forms of institutionalized racism has had enduring consequences for the economic well-being of African American, Latino, Native American, and other marginalized communities. Historic injustices such as the dispossession of land, exclusion from wealth-building opportunities, and systemic barriers to economic mobility have contributed to persistent disparities in wealth accumulation and intergenerational poverty.

Wealth Inequities: Racial disparities in wealth are stark and pervasive, with white households holding a disproportionate share of wealth compared to households of color. The racial wealth

53

gap reflects disparities in homeownership, access to financial assets, inheritance, and intergenerational wealth transfer. Structural barriers such as discriminatory lending practices, limited access to credit and capital, and disparities in homeownership rates contribute to the widening wealth gap between white households and communities of color.

Income Disparities: Racial disparities in income persist, with African American, Latino, and Native American workers earning lower wages and experiencing higher rates of unemployment and underemployment compared to their white counterparts. Factors such as occupational segregation, wage discrimination, and disparities in educational attainment contribute to income disparities and limit economic mobility for communities of color.

Educational Inequities: Disparities in educational opportunities and outcomes perpetuate racial inequalities in income, employment, and wealth accumulation. Persistent achievement gaps, unequal access to quality education, and disparities in school funding contribute to limited

economic opportunities and lower earning
potential for African American, Latino,
Native American, and other marginalized
students. Addressing educational
inequities and investing in equitable
access to quality education are essential
for narrowing the racial wealth gap and
promoting economic mobility.

Criminal Justice System: The
criminal justice system disproportionately
impacts communities of color, perpetuating
racial disparities in wealth, employment,
and housing. Racial profiling, disparities in
sentencing, mass incarceration, and
barriers to reentry exacerbate economic
inequalities and hinder opportunities for
upward mobility. Reforming the criminal
justice system, addressing systemic
biases, and investing in alternatives to
incarceration are critical for advancing
racial and economic justice.

Health Disparities: Racial disparities
in health outcomes and access to
healthcare contribute to the racial wealth
gap by imposing financial burdens and
limiting economic opportunities for
communities of color. Structural factors
such as unequal access to healthcare

services, environmental hazards, and disparities in health insurance coverage exacerbate health inequities and perpetuate economic disparities. Addressing health disparities and promoting equitable access to healthcare are essential components of efforts to reduce the racial wealth gap and promote economic equity.

Addressing racial disparities requires a holistic approach that addresses the root causes of systemic racism, dismantles structural barriers, and promotes policies and initiatives that advance racial and economic justice. By centering racial equity in efforts to address the wealth gap, policymakers, businesses, and communities can work together to build a more inclusive and equitable economy that benefits all Americans.

Educational Disparities

Educational disparities represent a significant contributor to the wealth gap in the United States, reflecting systemic inequities in access to quality education, resources, and opportunities. Disparities in educational attainment perpetuate socioeconomic inequalities, limiting economic mobility and contributing to the widening gap between affluent and marginalized communities. Understanding the complexities of educational disparities is crucial for devising effective strategies to address the wealth gap and promote equitable access to education for all Americans.

Unequal Access to Quality Education: Access to quality education is not equitable across communities, perpetuating disparities in academic achievement and economic opportunity. Low-income neighborhoods and communities of color often lack access to well-funded schools, experienced teachers, advanced coursework, and extracurricular activities, limiting educational opportunities and hindering academic success.

Resource Disparities: Disparities in school funding exacerbate educational inequalities, with affluent districts often receiving more resources and financial support than low-income and minority-majority schools. Disparities in funding levels, property tax revenues, and state funding formulas contribute to unequal access to educational resources,

57

technology, facilities, and support services, perpetuating academic achievement gaps and limiting opportunities for students in under-resourced schools.

Achievement Gaps:

Persistent achievement gaps persist along racial, socioeconomic, and geographic lines, reflecting disparities in academic performance, standardized test scores, graduation rates, and college readiness. Factors such as poverty, family background, language barriers, and inadequate support systems contribute to achievement gaps and limit educational attainment for students from disadvantaged backgrounds.

Segregation and School-to-Prison Pipeline: Segregation and inequities in school discipline policies contribute to the perpetuation of educational disparities, particularly for students of color and students with disabilities. The school-to-prison pipeline, characterized by harsh disciplinary practices, zero-tolerance policies, and disproportionate suspension and expulsion rates for marginalized students, exacerbates disparities in educational outcomes and perpetuates cycles of poverty and incarceration.

58

Higher Education Access and Affordability: Access to higher education remains a barrier for many low-income students and students of color, with rising tuition costs, student debt, and limited financial aid exacerbating disparities in college enrollment and completion rates. Limited access to college preparatory courses, standardized test preparation, and college advising services further hinder college access and perpetuate inequalities in educational attainment and economic mobility.

Digital Divide and Remote Learning: The COVID-19 pandemic has highlighted existing disparities in access to technology and broadband internet, exacerbating educational inequities for students from low-income households and rural communities. The digital divide limits access to online learning resources, virtual classrooms, and remote instruction, widening the gap in educational outcomes and exacerbating disparities in academic achievement and economic opportunity.

Addressing educational disparities requires a comprehensive approach that addresses the root causes of inequity, promotes

59

equitable funding and resources, and expands access to high-quality education for all students. By investing in early childhood education, supporting teacher professional development, expanding access to technology and broadband infrastructure, and dismantling systemic barriers to educational opportunity, policymakers, educators, and communities can work together to build a more equitable and inclusive education system that promotes economic mobility and reduces the wealth gap.

Systemic Barriers

Systemic barriers represent entrenched obstacles within societal structures and institutions that perpetuate the wealth gap and hinder economic mobility for marginalized communities. These barriers, rooted in historical injustices and structural inequalities, create unequal access to opportunities, resources, and social mobility, exacerbating disparities in wealth accumulation and perpetuating cycles of poverty and exclusion. Understanding the complexities of systemic barriers is essential for devising effective strategies to address the wealth gap and promote greater economic equity.

Discriminatory Practices:

Discriminatory practices within housing, employment, lending, and criminal justice systems perpetuate systemic inequalities and hinder economic advancement for marginalized communities. Racial discrimination in hiring, promotion, and wages limits employment opportunities and income potential for people of color, exacerbating disparities in wealth accumulation and economic mobility.

Unequal Access to Financial Services:

Limited access to banking services, credit, and capital disproportionately affects low-income communities and communities of color, hindering wealth-building opportunities and entrepreneurship. Discriminatory lending practices, redlining, and predatory financial products contribute to financial exclusion and perpetuate disparities in wealth accumulation and homeownership rates.

Health Disparities and Economic Burdens:

Health disparities, including unequal access to healthcare services, environmental hazards, and disparities in health outcomes, impose financial burdens and limit economic opportunities for marginalized communities. Lack of access to affordable healthcare, inadequate health insurance coverage, and disparities in health outcomes contribute to economic insecurity and perpetuate cycles of poverty and inequality.

Criminal Justice System: The criminal justice system disproportionately impacts communities of color, perpetuating racial disparities in wealth, employment, and housing. Racial profiling, disparities in sentencing, mass incarceration, and barriers to reentry exacerbate economic inequalities and hinder opportunities for upward mobility. Criminal records and involvement with the justice system can limit access to employment, housing, education, and financial services, perpetuating economic insecurity and hindering efforts to escape poverty.

Education and Opportunity Gaps: Educational inequities and opportunity gaps limit economic mobility and perpetuate intergenerational poverty. Disparities in school funding, resources,

and educational quality contribute to achievement gaps and limit opportunities for academic success and economic advancement. Limited access to college preparatory programs, advanced coursework, and extracurricular activities further exacerbate educational inequalities and perpetuate disparities in economic opportunity and wealth accumulation.

Environmental Injustice:

Environmental injustices, including exposure to pollution, environmental hazards, and lack of access to clean air and water, disproportionately affect low-income communities and communities of color. Environmental degradation and toxic exposure contribute to adverse health outcomes, economic burdens, and property devaluation, perpetuating disparities in wealth, health, and quality of life.

Addressing systemic barriers requires a comprehensive approach that addresses the root causes of inequity, promotes equitable policies and practices, and fosters greater inclusion and opportunity for marginalized communities. By dismantling discriminatory practices, investing in community development, expanding access to economic

63

opportunities and resources, and promoting policies that advance racial and economic justice, policymakers, businesses, and communities can work together to build a more equitable and inclusive economy that benefits all Americans.

Impact of the Wealth Gap on Society

The wealth gap in the United States exerts profound and far-reaching effects on society, touching every aspect of economic, social, and political life. Its impact reverberates across generations, perpetuating cycles of poverty, exacerbating social divisions, and undermining the principles of fairness, justice, and equality upon which our society is built. Understanding the consequences of the wealth gap is crucial for developing effective strategies to address inequality and foster a more equitable economy.

Economic Instability: The wealth gap undermines economic stability and resilience by concentrating wealth and economic power in the hands of a few, while leaving millions of Americans vulnerable to financial insecurity and hardship. Disparities in income, wealth, and access to economic opportunities limit consumer spending, dampen demand for goods and services, and hinder long-term economic growth and prosperity.

Social Cohesion: The widening wealth gap erodes social cohesion and trust within communities, fueling resentment, polarization, and social unrest. Growing disparities in wealth and opportunity exacerbate feelings of alienation and marginalization, undermining social bonds and eroding the sense of shared identity and common purpose essential for a healthy and cohesive society.

Political Polarization: Economic inequality contributes to political polarization and undermines democratic principles of representation and equality. The influence of money in politics, lobbying by special interests, and disparities in political participation amplify the voices of the wealthy and powerful at the expense of ordinary citizens, undermining democratic institutions and eroding public trust in government.

Inter-generational Mobility: The wealth gap perpetuates intergenerational

65

cycles of poverty and privilege, limiting opportunities for upward mobility and perpetuating disparities in economic opportunity and social mobility. Children born into poverty face systemic barriers to educational attainment, economic advancement, and access to resources and opportunities, perpetuating inequalities across generations and undermining the meritocratic ideals of American society.

Health Disparities: Economic inequality contributes to health disparities and undermines public health outcomes, with low-income individuals and communities experiencing higher rates of chronic illness, limited access to healthcare services, and poorer health outcomes compared to their wealthier counterparts. Inadequate access to healthcare, food insecurity, environmental hazards, and stress related to economic insecurity contribute to health inequities and perpetuate disparities in well-being.

Educational Inequities: The wealth gap contributes to educational inequities and undermines the potential of future generations. Disparities in educational opportunities, resources, and outcomes

66

limit access to quality education and perpetuate achievement gaps along racial, socioeconomic, and geographic lines, perpetuating cycles of poverty and limiting economic mobility and opportunity.

Innovation and Entrepreneurship:
Economic inequality stifles innovation and entrepreneurship by limiting access to capital, resources, and opportunities for aspiring entrepreneurs, particularly those from marginalized communities. Concentrated wealth and economic power in the hands of a few limit competition, innovation, and economic dynamism, hindering the potential for inclusive economic growth and prosperity.

Addressing the wealth gap requires bold and comprehensive solutions that address the root causes of inequality, dismantle systemic barriers to economic opportunity, and foster a more inclusive and equitable economy for all Americans. By investing in education, healthcare, affordable housing, and economic opportunity, promoting fair taxation and regulatory policies, and fostering a culture of inclusion and shared prosperity, we can build a society that values equity, justice, and opportunity for all.

Social Cohesion

Social cohesion, the degree of connectedness and solidarity within a society, is profoundly impacted by the wealth gap in the United States. As disparities in income, wealth, and opportunity widen, the fabric of society becomes strained, leading to fractures along socioeconomic lines and eroding the sense of shared identity and collective purpose. Understanding the dynamics of social cohesion is crucial for addressing the wealth gap and fostering a more inclusive and equitable society.

Erosion of Trust: The widening wealth gap undermines trust and social bonds within communities, fostering a sense of alienation and disconnection among individuals and groups. As disparities in wealth and opportunity deepen, feelings of resentment and distrust emerge, exacerbating social divisions and hindering collaboration and cooperation across lines of difference.

Polarization and Fragmentation: Economic inequality contributes to political polarization and social fragmentation, as individuals and groups retreat into ideological echo chambers and social bubbles. The concentration of wealth and power among the elite amplifies disparities in influence and access to resources, further fracturing society along lines of privilege and disadvantage.

Marginalization of Vulnerable Communities: The wealth gap disproportionately affects marginalized and vulnerable communities, exacerbating feelings of exclusion and marginalization. As disparities in wealth and opportunity widen, communities of color, low-income individuals, and other marginalized groups face systemic barriers to economic advancement and social inclusion, perpetuating cycles of poverty and exclusion.

Diminished Social Mobility: The wealth gap undermines social mobility and opportunity, limiting the ability of individuals and families to move up the economic ladder. As disparities in wealth and opportunity widen, the promise of upward mobility becomes increasingly elusive, leading to a sense of

disillusionment and resignation among those marginalized by the economic system.

Economic Segregation: Economic inequality contributes to spatial segregation and the stratification of communities along socioeconomic lines. As disparities in wealth and income deepen, affluent individuals and communities concentrate in enclaves of privilege, while low-income individuals and communities are relegated to neighborhoods with limited access to resources and opportunities, perpetuating cycles of segregation and inequality.

Civic Engagement and Participation: The widening wealth gap undermines civic engagement and participation, as marginalized communities face systemic barriers to political representation and advocacy. As disparities in wealth and opportunity deepen, marginalized individuals and communities are disenfranchised and disempowered, limiting their ability to shape public policy and advocate for change.

70

Social Solidarity and Resilience:

Despite the challenges posed by the wealth gap, social cohesion remains a powerful force for resilience and solidarity within communities. By fostering connections, building relationships, and nurturing empathy and understanding, communities can overcome divisions and work together to address the root causes of inequality and build a more just and equitable society. Addressing the wealth gap requires a concerted effort to promote social cohesion, foster inclusion, and build a sense of shared purpose and collective responsibility. By investing in social capital, strengthening social networks, and promoting dialogue and collaboration across lines of difference, we can bridge divides, foster solidarity, and build a society that values equity, justice, and opportunity for all.

Economic Growth

Economic growth serves as both a driver and a consequence of efforts to address the wealth gap in the United States. As the nation seeks to create a more equitable economy, it must recognize the role of economic growth in fostering prosperity,

71

reducing poverty, and promoting shared prosperity for all Americans. Understanding the dynamics of economic growth is essential for developing effective strategies to address the wealth gap and promote inclusive economic development.

Key Drivers of Economic Growth:

Economic growth is fueled by a combination of factors, including investment in physical and human capital, technological innovation, entrepreneurship, and productivity gains. Policies that promote innovation, encourage investment, foster entrepreneurship, and enhance productivity can stimulate economic growth and create opportunities for wealth creation and prosperity.

Reducing Poverty and Inequality: Economic growth has the potential to lift people out of poverty and reduce income inequality by creating jobs, increasing wages, and expanding opportunities for economic advancement. As the economy grows, individuals and families have greater access to employment, education, healthcare, and other essential services, reducing reliance on social assistance programs and promoting self-sufficiency and economic independence.

Expanding Access to Opportunity:

Economic growth opens doors to opportunity and mobility for individuals and communities, providing pathways to upward mobility and prosperity. As the economy expands, new opportunities emerge in emerging industries, growing sectors, and innovative enterprises, creating avenues for entrepreneurship, innovation, and economic advancement.

Investment in Human Capital:

Economic growth depends on investment in human capital, including education, training, and skills development. Policies that expand access to quality education, training programs, and lifelong learning opportunities can enhance human capital formation, increase workforce productivity, and promote economic growth and prosperity.

Inclusive Economic Development:

Sustainable economic growth must be inclusive, benefiting all segments of society and promoting shared prosperity. Inclusive economic development focuses on reducing disparities in income, wealth, and opportunity, ensuring that the benefits of growth are equitably distributed across communities and populations.

Environmental Sustainability:

Economic growth must be environmentally sustainable, balancing the need for economic development with the imperative to protect the planet and preserve natural resources for future generations. Policies that promote clean energy, sustainable infrastructure, and responsible resource management can support economic growth while safeguarding the environment and mitigating the impacts of climate change.

Social and Institutional Frameworks:

Economic growth is influenced by social and institutional frameworks that shape the business environment, promote competition, and ensure transparency, accountability, and the rule of law. Policies that strengthen institutions, promote good governance, and enhance the rule of law can create an enabling environment for economic growth and investment, fostering confidence and stability in the economy.

Globalization and Trade: Economic growth is increasingly interconnected with globalization and trade, as nations participate in global markets and supply chains. Policies that promote open trade, expand market access, and foster international cooperation can stimulate economic growth, create jobs, and promote prosperity both domestically and globally.

By recognizing the role of economic growth in addressing the wealth gap and promoting inclusive economic development, policymakers, businesses, and communities can work together to build a more equitable and prosperous economy that benefits all Americans. Through strategic investments, innovative policies, and collaborative efforts, we can harness the power of economic growth to create opportunities, reduce poverty, and promote shared prosperity for all.

Political Stability

Political stability is a cornerstone of a functioning democracy and plays a crucial role in addressing the wealth gap in the United States. A stable political environment fosters confidence, predictability, and trust in government institutions, creating an enabling environment for economic growth, social progress, and inclusive development. Understanding the importance of political stability is essential for devising effective strategies to address the wealth gap and promote a more equitable economy.

Confidence and Investment:
Political stability instills confidence among investors, businesses, and consumers, creating an environment conducive to investment, entrepreneurship, and economic growth. A stable political environment reduces uncertainty and risk, encouraging businesses to expand operations, invest in innovation, and create jobs, thereby stimulating economic activity and prosperity.

Policy Consistency and Predictability:
Political stability facilitates policy consistency and predictability, enabling governments to implement long-term strategies and reforms aimed at addressing the root causes of the wealth gap. Consistent and predictable policies provide clarity and

certainty to businesses and investors, encouraging them to make strategic investments and decisions that contribute to economic development and prosperity.

Social Cohesion and Unity:

Political stability fosters social cohesion and unity by promoting inclusive governance, dialogue, and collaboration among diverse stakeholders. A stable political environment encourages constructive engagement, compromise, and consensus-building, facilitating the resolution of differences and the pursuit of common goals and shared interests.

Democratic Institutions and Rule of Law:

Political stability is closely linked to the strength and resilience of democratic institutions and the rule of law. Strong democratic institutions, including independent judiciary, free press, and robust checks and balances, safeguard against abuses of power, corruption, and authoritarianism, ensuring accountability and transparency in government and promoting the rule of law.

Inclusive Governance and Participation:
Political stability requires inclusive governance structures that promote participation, representation, and empowerment of all segments of society, including marginalized and disadvantaged communities. Inclusive governance fosters trust, legitimacy, and social cohesion, ensuring that the benefits of political stability are equitably distributed across society and promoting a sense of belonging and ownership among citizens.

Civic Engagement and Democratic Values: Political stability encourages civic engagement, active citizenship, and the protection of democratic values and principles. A stable political environment enables citizens to participate in political processes, express their views, and hold government accountable for its actions, ensuring that policies and decisions reflect the interests and aspirations of the people.

Resilience to External Shocks:
Political stability enhances resilience to external shocks and crises, enabling governments and societies to respond effectively to challenges and disruptions. A

stable political environment fosters unity, solidarity, and cooperation in times of crisis, facilitating coordinated responses and collective action to mitigate risks and safeguard national interests.

International Standing and Reputation: Political stability enhances a nation's international standing and reputation, bolstering confidence among foreign investors, partners, and allies. A stable political environment signals strength, reliability, and credibility on the global stage, attracting investment, fostering trade relations, and promoting diplomatic cooperation and collaboration.

By prioritizing political stability and strengthening democratic institutions, governments, policymakers, and civil society can create an enabling environment for addressing the wealth gap and promoting inclusive economic development. Through inclusive governance, transparent decision-making, and effective leadership, we can build a more resilient, equitable, and prosperous society that benefits all Americans.

Case Studies and Data Analysis

Case Study: The Impact of Education on Economic Mobility: This case study examines the relationship between education and economic mobility by analyzing longitudinal data from diverse communities across the United States. Through in-depth interviews and quantitative analysis, the study explores how access to quality education, early childhood interventions, and post-secondary opportunities influence long-term economic outcomes and intergenerational wealth transfer. By highlighting success stories and best practices, the case study identifies strategies for improving educational equity and promoting upward mobility for disadvantaged populations.

Data Analysis: Racial Disparities in Wealth Accumulation: Using data from national surveys and longitudinal studies, this data analysis assesses racial disparities in wealth accumulation and explores the underlying factors driving these disparities. By disaggregating wealth data by race and ethnicity, the analysis uncovers persistent gaps in homeownership rates, retirement savings, and intergenerational wealth transfer between white households and communities of color. Through statistical modeling and regression analysis, the study identifies systemic barriers such as housing discrimination, employment disparities, and access to financial resources that contribute to racial inequalities in wealth accumulation.

Case Study: Community Development Initiatives and Economic Empowerment:

This case study examines the impact of community development initiatives and grassroots organizing on economic empowerment and wealth building in underserved communities. Drawing on qualitative interviews and participatory research methods, the study profiles successful community-driven projects, such as affordable housing developments, small business incubators, and cooperative enterprises, that have promoted economic self-sufficiency and social inclusion. By highlighting the role of community partnerships, local leadership, and collective action, the case study offers insights into effective strategies for addressing the wealth gap at the grassroots level.

Data Analysis: Gender Disparities in Wealth and Financial Security: This data analysis investigates gender disparities in wealth accumulation and financial security using national surveys and economic indicators. By examining differences in earnings, savings rates, investment patterns, and retirement savings between men and women, the analysis uncovers systemic barriers and structural inequalities that contribute to gender disparities in wealth. Through comparative analysis and gender-sensitive data modeling, the study identifies policy interventions and institutional reforms aimed at promoting gender equity and closing the wealth gap between genders.

Case Study: Impact Investing and Social Finance: This case study explores the role of impact investing and social finance in addressing the wealth gap and advancing inclusive economic development. Drawing on case studies from impact investors, philanthropic organizations, and social enterprises, the study examines innovative financing models and investment strategies that prioritize social impact alongside financial returns. By showcasing successful examples of impact investing in affordable housing, community development, and sustainable enterprises, the case study demonstrates how private capital can be leveraged to address systemic inequalities and promote shared prosperity.

Through rigorous data analysis and evidence-based case studies, this section of the book aims to deepen our understanding of the wealth gap and identify actionable strategies for creating a more equitable economy in the United States. By leveraging empirical research, qualitative insights, and real-world examples, policymakers, practitioners, and advocates can inform decision-making, drive policy change, and mobilize resources to address the root causes of economic inequality and promote inclusive economic growth.

CHAPTER 2

ROOT CAUSES OF THE WEALTH GAP

Historical Injustices: The wealth gap in the United States has deep historical roots, stemming from centuries of systemic injustices such as slavery, segregation, and discrimination. Historic injustices, including the dispossession of land, exclusion from wealth-building opportunities, and institutionalized racism, have had enduring consequences for the economic well-being of marginalized communities, perpetuating disparities in wealth accumulation and intergenerational poverty.

Structural Inequality: **Structural inequality embedded within economic, social, and political systems perpetuates the wealth gap by favoring the accumulation and preservation of wealth among the affluent while limiting opportunities for economic advancement among marginalized populations. Disparities in access to education, employment, housing, healthcare, and**

financial resources create barriers to economic mobility and perpetuate cycles of poverty and exclusion.

Wage Stagnation and Income Inequality: Wage stagnation and income inequality exacerbate the wealth gap by widening disparities in earnings and wages among individuals and households across the socioeconomic spectrum. Despite overall economic growth, wages for many workers have stagnated or grown at a slower pace compared to productivity gains and corporate profits, leading to a disproportionate concentration of wealth among the top income earners.

Racial and Gender Discrimination: Racial and gender discrimination within labor markets, housing markets, financial institutions, and educational systems contribute to disparities in wealth accumulation and economic opportunity. Persistent disparities in wages, employment, homeownership, access to credit, and investment opportunities limit economic mobility and perpetuate inequalities based on race, ethnicity, and gender.

Limited Access to Education and Opportunity: Limited access to quality education, workforce development programs, and economic opportunities exacerbates the wealth gap by perpetuating disparities in educational attainment, skills acquisition, and employment outcomes. Structural barriers such as inadequate school funding, disparities in educational resources, and systemic biases in hiring and promotion limit economic mobility and hinder the ability of individuals to build wealth and achieve financial security.

Unaffordable Housing and Rising Costs of Living: Unaffordable housing, rising costs of living, and gentrification contribute to the wealth gap by displacing low-income residents, limiting access to affordable housing, and exacerbating disparities in homeownership and property ownership. As housing costs escalate and income growth stagnates, low-income individuals and communities face increasing financial burdens and barriers to wealth accumulation.

Inequitable Tax Policies and Wealth Distribution: Inequitable tax policies and wealth distribution mechanisms exacerbate the wealth gap by favoring the wealthy and corporations at the expense of low-income and middle-income households. Tax cuts for high-income individuals, capital gains, and corporate profits disproportionately benefit the wealthiest segments of society, further widening income disparities and perpetuating inequalities in wealth accumulation and distribution.

Lack of Access to Financial Services and Investment Opportunities: Limited access to financial services, credit, and investment opportunities restricts economic mobility and wealth-building potential for low-income individuals and communities. Discriminatory lending practices, redlining, and predatory financial products create barriers to financial inclusion and perpetuate disparities in access to capital, investment opportunities, and wealth-building resources.

87

Addressing the root causes of the wealth gap requires a comprehensive approach that tackles systemic inequalities, dismantles structural barriers, and promotes policies and initiatives that foster economic opportunity, equity, and inclusion for all Americans. By addressing historical injustices, promoting equitable policies, and investing in education, workforce development, affordable housing, and financial empowerment, policymakers, businesses, and communities can work together to create a more equitable and inclusive economy that benefits everyone.

Economic Policies and Structures

Progressive Taxation: Implementing progressive taxation policies that place a higher tax burden on the wealthy can help redistribute wealth and reduce income inequality. Progressive taxation can include higher marginal tax rates for high-income earners, capital gains taxes, estate taxes, and corporate taxes. Revenue generated from progressive taxation can be used to fund social programs, education, healthcare, and infrastructure investments that benefit low- and middle-income families.

Minimum Wage Increases:
Increasing the minimum wage to a living wage can lift millions of workers out of poverty and narrow the wealth gap. Indexing the minimum wage to inflation ensures that workers' purchasing power keeps pace with the cost of living. Additionally, implementing policies that promote wage transparency, pay equity, and workplace protections can help address disparities in wages and promote economic security for low-wage workers.

Investments in Education and Workforce Development:
Investing in education and workforce development programs can provide individuals with the skills, training, and credentials needed to compete in the 21st-century economy. Increasing funding for public education, expanding access to vocational training, apprenticeships, and lifelong learning opportunities can improve educational outcomes, increase earning potential, and promote economic mobility for disadvantaged populations.

89

Affordable Housing Initiatives:
Implementing affordable housing initiatives, such as subsidized housing, rent control measures, and housing vouchers, can address the housing affordability crisis and promote equitable access to safe, stable, and affordable housing. Investing in affordable housing development, preserving existing affordable housing units, and implementing anti-displacement policies can help prevent homelessness, reduce housing segregation, and promote inclusive communities.

Financial Inclusion and Access to Banking Services:
Expanding access to banking services, credit, and financial literacy programs can empower low-income individuals and communities to build assets, establish credit histories, and access capital for investment and entrepreneurship. Implementing policies that promote financial inclusion, such as community banking initiatives, microfinance programs, and alternative lending models, can help bridge the wealth gap and promote economic empowerment for underserved populations.

Job Creation and Economic Stimulus:

Implementing targeted job creation programs and economic stimulus measures can stimulate economic growth, create employment opportunities, and reduce poverty. Investing in infrastructure projects, green energy initiatives, and small business development programs can generate jobs, spur innovation, and promote inclusive economic development in underserved communities.

Worker Protections and Labor Rights:

Strengthening worker protections, enforcing labor standards, and promoting collective bargaining rights can empower workers to advocate for fair wages, safe working conditions, and benefits. Implementing policies that protect workers from exploitation, discrimination, and wage theft, such as paid sick leave, family leave, and overtime protections, can promote economic security and reduce income inequality.

Corporate Accountability and Responsible Business Practices:

Holding corporations accountable for their social and environmental impacts can promote responsible business practices and corporate citizenship. Implementing policies that require corporations to disclose their environmental, social, and governance practices, promote diversity and inclusion in the workplace, and invest in sustainable business practices can foster greater transparency, accountability, and ethical behavior in the private sector.

By implementing these and other progressive economic policies and structures, policymakers, businesses, and communities can work together to address the wealth gap, promote economic equity, and build a more inclusive and sustainable economy that benefits all Americans. Through bold and visionary leadership, collective action, and a commitment to social justice, we can create a future where prosperity is shared, opportunities are accessible, and every individual has the chance to thrive.

Systemic Racism and Discrimination

Systemic racism and discrimination represent significant barriers to economic opportunity and wealth accumulation for communities of color in the United States.

92

Rooted in historical injustices and perpetuated by structural inequalities within economic, social, and political systems, systemic racism and discrimination contribute to the widening wealth gap and perpetuate cycles of poverty and exclusion. Understanding the impact of systemic racism and discrimination is essential for devising effective strategies to address the wealth gap and promote racial equity in the economy.

Historical Context: Systemic racism and discrimination have deep historical roots in the United States, dating back to slavery, segregation, and institutionalized racism. Historic injustices, including the legacy of slavery, Jim Crow laws, redlining, and discriminatory housing policies, have had enduring consequences for the economic well-being of African American, Latino, Native American, and other marginalized communities, limiting access to wealth-building opportunities and perpetuating intergenerational poverty.

Disparities in Employment and Income: Systemic racism and discrimination contribute to disparities in

93

employment, wages, and income between white Americans and communities of color. Persistent disparities in hiring, promotion, and pay practices, along with occupational segregation and wage gaps, limit economic mobility and hinder opportunities for wealth accumulation among marginalized populations. Discriminatory practices such as racial profiling, workplace harassment, and biased hiring decisions further exacerbate inequalities in the labor market.

Access to Education and Opportunity: Systemic racism and discrimination create barriers to educational attainment and economic opportunity for communities of color. Disparities in access to quality education, inadequate funding for schools in low-income neighborhoods, and discriminatory disciplinary practices contribute to educational inequities and limit economic mobility for African American, Latino, Native American, and other minority students. Limited access to college preparatory programs, advanced coursework, and extracurricular activities further perpetuate disparities in

educational outcomes and economic opportunity.

Housing Discrimination and Segregation:

Systemic racism and discrimination in housing markets perpetuate disparities in homeownership rates, property values, and access to affordable housing for communities of color. Historic practices such as redlining, restrictive covenants, and predatory lending have systematically excluded African American and other minority communities from accessing homeownership and building wealth through property ownership. Persistent housing discrimination, gentrification, and displacement further exacerbate inequalities in housing and contribute to the wealth gap.

Health Disparities and Economic Burdens:

Systemic racism and discrimination contribute to health disparities and economic burdens for communities of color. Limited access to healthcare services, disparities in health outcomes, and environmental injustices disproportionately affect African American, Latino, Native American, and other

95

marginalized communities, leading to higher rates of chronic illness, premature mortality, and economic hardship. Inadequate health insurance coverage, limited access to preventive care, and discriminatory healthcare practices exacerbate economic insecurity and perpetuate disparities in wealth and well-being.

Criminal Justice System Inequities:

Systemic racism and discrimination within the criminal justice system contribute to disparities in incarceration rates, sentencing outcomes, and economic opportunities for communities of color. Racial profiling, biased policing practices, disparities in bail and sentencing, and barriers to reentry create cycles of poverty and incarceration that disproportionately affect African American, Latino, and Native American individuals and families. Criminal records and involvement with the justice system limit access to employment, housing, education, and financial services, perpetuating economic insecurity and hindering efforts to escape poverty.

Addressing systemic racism and discrimination requires a multifaceted approach that addresses the root causes of inequality, dismantles structural barriers, and promotes racial equity and inclusion in all aspects of society. By implementing policies that combat racial disparities in employment, education, housing, healthcare, and criminal justice, policymakers, businesses, and communities can work together to build a more equitable and inclusive economy that promotes shared prosperity and opportunity for all Americans. Through intentional efforts to address systemic racism and discrimination, we can create a future where every individual has the opportunity to thrive and contribute to the prosperity of our nation.

Educational and Employment Systems

Equitable Education Funding: Implementing equitable education funding mechanisms that allocate resources based on student needs and school district characteristics can help address disparities in educational quality and opportunity. Fair funding formulas that account for factors such as poverty levels, English language learner populations, and special education needs can ensure that all students have access to high-quality education regardless of their socioeconomic background or zip code.

Investment in Early Childhood Education: Investing in early childhood education programs, such as universal pre-K, Head Start, and early intervention services, can help narrow the opportunity gap and promote long-term academic success and economic mobility. Providing access to high-quality early childhood education for all children, especially those from low-income and disadvantaged backgrounds, can improve school readiness, reduce achievement gaps, and increase graduation rates.

College Affordability and Student Debt Relief: Addressing college affordability and student debt can remove barriers to higher education and promote economic opportunity for low-income and minority students. Implementing policies such as tuition-free college programs, need-based financial aid, and student loan forgiveness programs can make higher education more accessible and affordable for all students, regardless of their socioeconomic status or family background.

Promoting Career and Technical Education (CTE):

Expanding access to career and technical education (CTE) programs and apprenticeship opportunities can provide students with the skills, training, and credentials needed to succeed in high-demand industries and occupations. By aligning CTE programs with local workforce needs and industry certifications, students can gain valuable skills and experience that lead to well-paying jobs and career advancement opportunities.

Workforce Development and Lifelong Learning:

Investing in workforce development programs and lifelong learning opportunities can help workers adapt to changing economic conditions and technological advancements. Providing access to job training, upskilling programs, and adult education initiatives can empower workers to acquire new skills, transition to new

99

industries, and remain competitive in the labor market throughout their careers.

Diversity and Inclusion Initiatives: Promoting diversity and inclusion within educational institutions and workplaces can foster a more equitable and inclusive society. Implementing policies that promote diversity in hiring, admissions, and promotion decisions, as well as fostering inclusive environments that value diverse perspectives and experiences, can help dismantle systemic barriers and promote equal opportunity for all individuals regardless of their race, ethnicity, gender, or socioeconomic status.

Career Pathways and Mentorship Programs: Creating clear career pathways and mentorship programs can help students and workers navigate their educational and career journeys and access opportunities for advancement. Providing mentorship, guidance, and support to students and workers from underrepresented backgrounds can help break down barriers, build social capital, and promote upward mobility in the workforce.

100

Promoting Entrepreneurship and Small Business Development:

Supporting entrepreneurship and small business development can create opportunities for wealth creation and economic empowerment, particularly for minority-owned businesses and communities. Providing access to capital, technical assistance, and business development resources can help aspiring entrepreneurs overcome barriers to entry, launch successful businesses, and contribute to local economic development and job creation.

By addressing structural inequalities within educational and employment systems and promoting policies and initiatives that foster equity, inclusion, and opportunity for all individuals, policymakers, educators, employers, and community leaders can work together to create a more equitable and prosperous society where every individual has the opportunity to succeed and thrive. Through collective efforts to address the root causes of the wealth gap, we can build a future where economic opportunity is accessible to all Americans,

101

regardless of their background or circumstances.

Intergenerational Wealth Transfer

Intergenerational wealth transfer plays a significant role in perpetuating the wealth gap in the United States. Passed down from one generation to the next, wealth accumulated over time can serve as a source of economic stability, opportunity, and social mobility for some families while perpetuating disparities in wealth and opportunity for others. Understanding the dynamics of intergenerational wealth transfer is essential for addressing the wealth gap and promoting economic equity and inclusion.

Historical Accumulation of Wealth: Intergenerational wealth transfer often reflects historical patterns of wealth accumulation and privilege, shaped by factors such as inheritance, property ownership, business ownership, and access to financial resources. Families with a history of wealth accumulation and asset ownership are better positioned to pass down wealth to future generations,

creating advantages and opportunities that perpetuate disparities in wealth and economic opportunity.

Inheritance and Estate Planning:
Inheritance and estate planning play a central role in intergenerational wealth transfer, allowing families to pass down assets, property, investments, and business interests to heirs and beneficiaries. Inherited wealth can provide financial security, access to education, homeownership opportunities, and entrepreneurial capital for future generations, contributing to the perpetuation of economic advantages and disparities across families and communities.

Disparities in Inheritance and Wealth Distribution:
Disparities in inheritance and wealth distribution reflect broader inequalities in access to intergenerational wealth transfer mechanisms and estate planning strategies. Families with higher levels of wealth and assets are more likely to engage in estate planning, trust formation, and tax optimization strategies to

103

maximize the transfer of wealth to heirs and beneficiaries, exacerbating disparities in inherited wealth and economic opportunity across socioeconomic and demographic groups.

Impact on Economic Mobility:

Intergenerational wealth transfer has a significant impact on economic mobility and opportunity, shaping the life chances and opportunities available to individuals and families across generations. Inherited wealth can provide a financial safety net, access to educational opportunities, and capital for entrepreneurship and investment, facilitating upward mobility and economic advancement for some individuals while perpetuating barriers to wealth accumulation and economic mobility for others.

Inequality of Opportunity:

Intergenerational wealth transfer can reinforce inequality of opportunity by perpetuating advantages and disadvantages across generations. Families with access to intergenerational wealth transfer mechanisms and inheritance strategies are better positioned to accumulate assets, build financial security, and access opportunities for economic advancement, while those without access to inherited wealth face greater challenges in overcoming barriers to wealth accumulation and achieving economic mobility.

Policy Implications and Solutions:
Addressing the wealth gap requires policies and interventions that promote equitable access to intergenerational wealth transfer mechanisms and opportunities for wealth accumulation. Policies such as estate tax reform, wealth taxation, inheritance tax reform, and asset-building initiatives can help mitigate disparities in inherited wealth and promote greater economic equity and inclusion. Additionally, investing in education, workforce development, affordable housing, and financial empowerment programs can create pathways to economic opportunity and mobility for individuals and families across generations.

Promoting Financial Literacy and Wealth Education:
Promoting financial literacy, wealth education, and asset-building strategies can empower

105

individuals and families to make informed decisions about wealth management, estate planning, and intergenerational wealth transfer. Providing access to financial education resources, estate planning services, and asset-building programs can help individuals and families build financial resilience, navigate intergenerational wealth transfer processes, and maximize opportunities for economic advancement and security.

By addressing the dynamics of intergenerational wealth transfer and promoting policies and initiatives that foster economic equity and inclusion, policymakers, businesses, and communities can work together to create a more equitable and prosperous society where every individual has the opportunity to build wealth, achieve economic security, and realize their full potential. Through intentional efforts to address the root causes of the wealth gap, we can build a future where economic opportunity is accessible to all Americans, regardless of their background or circumstances.

CHAPTER 3

SOLUTIONS FOR ADDRESSING THE WEALTH GAP

Progressive Taxation Policies: Implementing progressive taxation policies that place a higher tax burden on the wealthy can help redistribute wealth and reduce income inequality. By increasing taxes on high-income earners, capital gains, and corporate profits, governments can generate revenue to fund social programs, education, healthcare, and infrastructure investments that benefit low- and middle-income families.

Raise the Minimum Wage: Increasing the minimum wage to a living wage can lift millions of workers out of poverty and narrow the wealth gap. Indexing the minimum wage to inflation ensures that workers' purchasing power keeps pace with the cost of living. Additionally, policies that promote wage transparency, pay equity, and workplace protections can help address disparities in wages and promote economic security for low-wage workers.

Investments in Education and Workforce Development: Investing in education and workforce development programs can provide individuals with the skills, training, and credentials needed to compete in the 21st-century economy. Increasing funding for public education, expanding access to vocational training, apprenticeships, and lifelong learning opportunities can improve educational outcomes, increase workforce participation, and promote economic mobility for disadvantaged populations.

Affordable Housing Initiatives: Implementing affordable housing initiatives, such as subsidized housing, rent control measures, and housing vouchers, can address the housing affordability crisis and promote equitable access to safe, stable, and affordable housing. Investing in affordable housing development, preserving existing affordable housing units, and implementing anti-displacement policies can help prevent homelessness, reduce housing segregation, and promote inclusive communities.

Financial Inclusion and Access to Banking Services:

Expanding access to banking services, credit, and financial literacy programs can empower low-income individuals and communities to build assets, establish credit histories, and access capital for investment and entrepreneurship. Implementing policies that promote financial inclusion, such as community banking initiatives, microfinance programs, and alternative lending models, can help bridge the wealth gap and promote economic empowerment for underserved populations.

Promote Economic Opportunity Zones:

Creating economic opportunity zones in distressed communities can attract investment, spur economic development, and create jobs in underserved areas. Offering tax incentives, grants, and technical assistance to businesses and investors operating in opportunity zones can revitalize local economies, promote entrepreneurship, and expand economic opportunities for residents.

Combat Systemic Racism and Discrimination:

Addressing systemic racism and discrimination within economic, social, and political systems is essential for closing the wealth gap and promoting racial equity. Implementing policies that promote diversity and inclusion in hiring, education, housing, and financial services, as well as addressing disparities in healthcare, criminal justice, and environmental justice, can help dismantle structural barriers and create pathways to economic opportunity for marginalized communities.

Promote Entrepreneurship and Small Business Development:

Supporting entrepreneurship and small business development can create opportunities for wealth creation and economic empowerment, particularly for minority-owned businesses and communities. Providing access to capital, technical assistance, and business development resources can help aspiring entrepreneurs overcome barriers to entry, launch successful businesses, and contribute to local economic development and job creation.

110

Community Wealth-Building Initiatives:

Implementing community wealth-building initiatives, such as community land trusts, worker cooperatives, and community development financial institutions (CDFIs), can empower communities to collectively own and control assets, generate wealth, and address local economic challenges. By promoting community ownership, democratic decision-making, and equitable development, these initiatives can help build resilience, promote economic self-sufficiency, and reduce dependence on external sources of capital and investment.

Promote Financial Literacy and Wealth Education:

Promoting financial literacy, wealth education, and asset-building strategies can empower individuals and families to make informed decisions about wealth management, estate planning, and intergenerational wealth transfer. Providing access to financial education resources, estate planning services, and asset-building programs can help individuals and families build financial resilience, navigate intergenerational wealth transfer

111

processes, and maximize opportunities for economic advancement and security.

By implementing these and other proactive solutions, policymakers, businesses, and communities can work together to address the root causes of the wealth gap, promote economic equity and inclusion, and build a more prosperous and equitable society where everyone has the opportunity to thrive and succeed. Through collective action and a commitment to social justice, we can create a future where economic opportunity is accessible to all Americans, regardless of their background or circumstances.

Policy Reforms for Addressing the Wealth Gap

Wealth Taxation: Implementing a wealth tax on high-net-worth individuals and households can help address wealth concentration and reduce economic inequality. A wealth tax levies an annual tax on the net worth of individuals above a certain threshold, thereby redistributing wealth and generating revenue for social programs, education, healthcare, and infrastructure investments that benefit low- and middle-income families.

Estate Tax Reform: Reforming the estate tax can prevent the accumulation of dynastic wealth and promote greater intergenerational equity. Adjusting estate tax rates, lowering exemption thresholds, and closing loopholes can ensure that wealth is distributed more equitably across generations and reduce disparities in inherited wealth between affluent families and the broader population.

Corporate Tax Reform: Reforming corporate taxation can ensure that corporations pay their fair share of taxes and contribute to addressing the wealth gap. Closing corporate tax loopholes, implementing a minimum corporate tax rate, and taxing corporate profits and offshore earnings can generate revenue for public investments and social programs that promote economic opportunity and mobility for all Americans.

Financial Regulation and Consumer Protection: Strengthening financial regulation and consumer protection laws can safeguard against predatory lending practices, abusive financial products, and discriminatory lending practices that perpetuate

113

economic inequality. Implementing stricter regulations on payday lenders, debt collectors, and subprime lenders, as well as enforcing fair lending laws and anti-discrimination statutes, can protect consumers and promote financial inclusion and equity.

Universal Basic Income (UBI):

Implementing a universal basic income program can provide all citizens with a guaranteed income floor, ensuring that everyone has access to basic necessities and economic security. A universal basic income can reduce poverty, inequality, and economic insecurity, while promoting individual autonomy, dignity, and freedom of choice in how individuals pursue their economic aspirations and goals.

Guaranteed Jobs Programs:

Establishing guaranteed jobs programs can ensure that everyone who wants to work has access to meaningful employment opportunities that provide fair wages, benefits, and workplace protections. Creating public-sector jobs in areas such as infrastructure, renewable energy, education, healthcare, and community development can stimulate economic growth, reduce unemployment, and address structural inequalities in the labor market.

Universal Healthcare Coverage:
Implementing universal healthcare coverage can reduce financial barriers to healthcare access and promote health equity for all Americans. Establishing a single-payer healthcare system or expanding public health insurance programs can ensure that everyone has access to affordable, comprehensive healthcare services, regardless of their income, employment status, or pre-existing conditions.

Housing Policy Reforms: Reforming
housing policies can increase access to affordable housing and reduce housing insecurity for low- and moderate-income families. Implementing rent control measures, expanding housing subsidies, and investing in affordable housing development can address housing affordability challenges, reduce homelessness, and promote stable and inclusive communities.

Education Funding Equity:
Ensuring equitable funding for public

115

education can promote educational equity and reduce disparities in educational outcomes and opportunities. Implementing state funding formulas that allocate resources based on student needs, reducing reliance on property taxes for school funding, and increasing investments in high-poverty schools can improve educational quality, narrow achievement gaps, and promote social mobility for disadvantaged students.

Criminal Justice Reform:

Implementing criminal justice reforms can address systemic inequalities in the criminal justice system and reduce barriers to economic opportunity and mobility for individuals with criminal records. Revising sentencing laws, promoting alternatives to incarceration, expunging non-violent criminal records, and investing in reentry programs and rehabilitation services can facilitate successful reintegration into society and promote economic inclusion for justice-involved individuals.

By advocating for and implementing these policy reforms, policymakers, advocates, and community leaders can work together to address the root causes of the wealth gap, promote economic equity and

116

inclusion, and build a more prosperous and equitable society for all Americans. Through collective action and a commitment to social justice, we can create a future where economic opportunity is accessible to everyone, regardless of their background or circumstances.

Taxation Policies for Addressing the Wealth Gap

Progressive Income Tax: Implementing a progressive income tax system can help address income inequality by taxing higher incomes at higher rates. Progressive taxation ensures that those who can afford to contribute more to society do so, while providing relief for lower-income individuals and families. Adjusting tax brackets and rates to ensure that the wealthiest individuals pay their fair share can generate revenue for social programs and investments that benefit the broader population.

Capital Gains Tax Reform: Reforming the capital gains tax can help reduce wealth concentration and address disparities in investment income. Taxing

117

capital gains at rates closer to ordinary income rates, particularly for high-income earners, can prevent wealthy individuals from accumulating wealth through investments at lower tax rates than those applied to wages and salaries. Closing loopholes and implementing measures to prevent tax evasion can ensure that investment income is subject to fair taxation.

Inheritance and Estate Taxes:
Strengthening inheritance and estate taxes can prevent the concentration of wealth across generations and promote greater intergenerational equity. Increasing estate tax rates, lowering exemption thresholds, and closing loopholes can ensure that inherited wealth is distributed more equitably and used to fund public investments and social programs that benefit the broader population. Implementing progressive estate tax rates based on the size of the estate can prevent dynastic wealth accumulation and promote economic mobility.

Wealth Taxation: Introducing a
wealth tax on high-net-worth individuals and households can help address wealth

concentration and reduce economic inequality. A wealth tax levies an annual tax on the net worth of individuals above a certain threshold, redistributing wealth and generating revenue for investments in education, healthcare, infrastructure, and other initiatives that promote economic opportunity and mobility. Implementing a progressive wealth tax system that applies higher rates to larger fortunes can ensure that the wealthiest individuals contribute their fair share to society.

Corporate Taxation Reform:
Reforming corporate taxation can ensure that corporations pay their fair share of taxes and contribute to addressing the wealth gap. Closing corporate tax loopholes, implementing a minimum corporate tax rate, and taxing corporate profits and offshore earnings can generate revenue for public investments and social programs that promote economic opportunity and mobility for all Americans. Requiring corporations to disclose their tax practices and pay their fair share can promote transparency and accountability in the corporate sector.

119

Financial Transaction Taxes: Implementing a financial transaction tax on stock trades, bond sales, and other financial transactions can generate revenue while discouraging speculative trading and excessive risk-taking in financial markets. A financial transaction tax can help curb financial market volatility and excessive speculation while generating revenue for investments in infrastructure, education, healthcare, and other public goods that benefit the broader population.

Closing Tax Loopholes and Offshore Tax Havens: Closing tax loopholes and offshore tax havens can prevent wealthy individuals and corporations from avoiding taxation and shifting their tax burdens onto ordinary taxpayers. Implementing measures to prevent tax evasion, closing offshore tax shelters, and requiring greater transparency in financial transactions can ensure that everyone pays their fair share of taxes and contribute to reducing the wealth gap.

By implementing progressive taxation policies that promote fairness, equity, and

120

transparency, policymakers can address the root causes of the wealth gap and promote economic opportunity and mobility for all Americans. Through collective action and a commitment to social justice, we can create a tax system that fosters shared prosperity and opportunity for generations to come.

Labor Laws and Minimum Wage

Minimum Wage Increase: Raising the minimum wage to a living wage is crucial for addressing the wealth gap and promoting economic equity. A living wage ensures that workers earn enough income to cover basic living expenses such as housing, food, healthcare, and transportation. Increasing the minimum wage not only lifts workers out of poverty but also stimulates consumer spending, reduces income inequality, and promotes economic growth.

Indexing the Minimum Wage to Inflation: Indexing the minimum wage to inflation ensures that wages keep pace with the rising cost of living. By adjusting the minimum wage annually based on changes in the Consumer Price Index (CPI) or other inflation indicators, policymakers can prevent erosion in the purchasing power of low-wage workers and ensure that wages reflect changes in economic conditions over time.

Elimination of Subminimum Wage for Tipped Workers and Disabled Workers: Eliminating the subminimum wage for tipped workers and disabled workers is essential for promoting fair labor practices and economic equity. Subminimum wage laws allow employers to pay tipped workers and disabled workers below the standard minimum wage, leading to financial insecurity and exploitation. Phasing out subminimum wage provisions and ensuring that all workers receive a living wage can help close the wage gap and improve economic security for vulnerable workers.

Equal Pay Legislation: Enacting equal pay legislation to address gender and racial wage gaps is critical for promoting economic equality and closing the wealth gap. Equal pay laws require employers to provide equal compensation for equal work, regardless of gender, race, ethnicity, or other protected characteristics. Strengthening enforcement mechanisms and penalties for pay discrimination can help ensure that all workers receive fair and equitable compensation for their contributions.

Fair Labor Standards Act (FLSA) Enforcement: Strengthening enforcement of the Fair Labor Standards Act (FLSA) is essential for protecting workers' rights and ensuring compliance with minimum wage, overtime, and child labor provisions. Increasing funding for wage and hour enforcement agencies, expanding outreach and education efforts, and implementing stricter penalties for wage theft and labor violations can deter employers from engaging in exploitative practices and protect workers from economic exploitation.

Expansion of Collective Bargaining Rights:
Expanding collective bargaining rights for workers empowers employees to negotiate for better wages, benefits, and working conditions. Strengthening protections for workers' rights to organize and bargain collectively, removing barriers to

unionization, and promoting sectoral bargaining arrangements can help rebalance power dynamics in the labor market and ensure that workers have a voice in shaping their workplaces and livelihoods.

Guaranteed Paid Sick Leave and Family Leave: Guaranteeing paid sick leave and family leave for all workers is essential for promoting economic security and work-life balance. Paid leave policies enable workers to take time off to address personal or family health needs without sacrificing income or job security. Implementing universal paid sick leave and family leave programs can reduce financial hardship, improve public health outcomes, and support working families in achieving economic stability and well-being.

Protection of Gig Workers and Independent Contractors: Extending labor protections and benefits to gig workers and independent contractors is critical for ensuring that all workers have access to basic labor rights and protections. Reforming labor laws to recognize gig workers as employees, expanding access to unemployment

124

insurance, healthcare benefits, and retirement savings programs, and establishing portable benefits models can provide greater economic security and stability for workers in the gig economy.

By enacting labor laws and policies that promote fair wages, protect workers' rights, and ensure economic security for all individuals, policymakers can address the root causes of the wealth gap and create a more equitable economy that benefits everyone. Through collective action and a commitment to social justice, we can build a future where work is valued, wages are fair, and opportunities for economic advancement are accessible to all Americans.

Social Welfare Programs for Addressing the Wealth Gap

Universal Healthcare Coverage:

Implementing universal healthcare coverage ensures that all Americans have access to affordable and comprehensive healthcare services, regardless of their income, employment status, or pre-existing conditions. A single-payer healthcare system or public option model can reduce healthcare costs, improve health outcomes, and alleviate financial burdens associated with medical expenses, thereby promoting economic security and well-being for individuals and families.

Expanded Medicaid and Medicare: Expanding Medicaid eligibility and strengthening Medicare benefits can provide vital healthcare coverage to low-income individuals, seniors, and people with disabilities. Increasing funding for Medicaid expansion initiatives, closing the Medicaid coverage gap, and enhancing Medicare benefits for preventive care, prescription drugs, and long-term care services can improve health equity and reduce disparities in access to healthcare services among vulnerable populations.

Supplemental Nutrition Assistance Program (SNAP): Strengthening and expanding the Supplemental Nutrition Assistance Program (SNAP) ensures that low-income individuals and families have access to nutritious food and essential nutrition assistance. Increasing SNAP benefits, expanding eligibility criteria, and investing in nutrition education and outreach programs can alleviate food insecurity, improve dietary

126

quality, and promote health and well-being for millions of Americans struggling with poverty and hunger.

Affordable Housing Initiatives:

Investing in affordable housing initiatives, such as subsidized housing, rental assistance programs, and homelessness prevention services, can address housing affordability challenges and reduce homelessness among low-income individuals and families. Increasing funding for affordable housing development, expanding rental assistance vouchers, and implementing tenant protections can ensure that everyone has access to safe, stable, and affordable housing, thereby promoting economic stability and housing security.

Unemployment Insurance (UI):

Strengthening and modernizing the unemployment insurance (UI) system provides critical financial support to workers who lose their jobs through no fault of their own. Expanding UI eligibility, increasing benefit levels, and extending the duration of benefits during economic downturns can help stabilize incomes, prevent poverty, and support individuals and families during periods of unemployment or underemployment.

Earned Income Tax Credit (EITC): Expanding the Earned Income Tax Credit (EITC) provides financial assistance to low- and moderate-income workers and families, lifting millions of people out of poverty each year. Increasing the EITC benefit amounts, expanding eligibility criteria, and making the credit fully refundable can boost incomes, incentivize work, and reduce poverty rates among working families struggling to make ends meet.

Childcare Assistance Programs: Investing in childcare assistance programs, such as subsidized childcare, early childhood education, and afterschool programs, supports working parents and promotes children's development and school readiness. Increasing funding for childcare subsidies, expanding access to high-quality childcare providers, and implementing policies that improve childcare affordability and accessibility can help families balance work and caregiving responsibilities, support women's workforce participation, and

128

promote child well-being and school success.

Education and Training Programs: Investing in education and training programs, such as Pell Grants, workforce development initiatives, and adult education programs, can provide individuals with the skills, training, and credentials needed to succeed in the labor market and pursue higher-paying jobs. Increasing funding for education and training programs, expanding access to tuition assistance and career counseling services, and promoting pathways to college and career readiness can empower individuals to achieve economic mobility and build brighter futures for themselves and their families.

Disability Benefits and Support Services: Strengthening disability benefits and support services ensures that individuals with disabilities have access to essential financial assistance, healthcare coverage, and supportive services that enable them to live with dignity and independence. Expanding access to Social Security Disability Insurance (SSDI) and Supplemental Security Income (SSI),

129

improving access to healthcare and assistive technologies, and enhancing vocational rehabilitation and employment services can promote economic security and inclusion for people with disabilities.

Community Development Programs: Investing in community development programs, such as community economic development initiatives, small business assistance programs, and neighborhood revitalization efforts, fosters economic growth, creates jobs, and builds stronger and more resilient communities. Supporting community-based organizations, minority-owned businesses, and local entrepreneurship initiatives can promote economic empowerment, address disparities in access to capital and resources, and stimulate inclusive economic development in underserved communities.

By strengthening social welfare programs and safety net policies, policymakers can address the root causes of the wealth gap, promote economic opportunity and mobility, and build a more equitable and inclusive society where everyone has the opportunity to thrive and succeed. Through

130

collective action and a commitment to social justice, we can create a future where economic security, health, and well-being are accessible to all Americans, regardless of their background or circumstances.

Affordable Housing Initiatives

Subsidized Housing Programs:
Implementing subsidized housing programs provides affordable housing options for low-income individuals and families. These programs, such as public housing and Section 8 vouchers, offer rental assistance to eligible households, enabling them to access safe and decent housing at below-market rates. Increasing funding for subsidized housing programs, expanding eligibility criteria, and reducing waitlists can address housing affordability challenges and reduce homelessness among vulnerable populations.

Low-Income Housing Tax Credits (LIHTC):
Utilizing Low-Income Housing Tax Credits (LIHTC) incentivizes private developers to invest in the construction or rehabilitation of affordable

131

housing units for low-income residents.
LIHTC programs offer tax credits to
developers in exchange for providing
affordable housing units that meet income
eligibility requirements. Expanding LIHTC
allocations, streamlining application
processes, and prioritizing affordable
housing projects in underserved
communities can increase the supply of
affordable housing and address housing
shortages in high-cost markets.

Affordable Housing Trust Funds:

Establishing affordable housing trust funds
at the local, state, and federal levels
provides dedicated funding for affordable
housing development and preservation
initiatives. Affordable housing trust funds
allocate resources for land acquisition,
construction, rehabilitation, and rental
assistance programs that prioritize housing
affordability and accessibility for low- and
moderate-income households.
Implementing dedicated funding sources,
such as real estate transfer taxes or
developer impact fees, can ensure
sustainable funding for affordable housing
initiatives and promote equitable
development.

132

Inclusionary Zoning Policies:

Implementing inclusionary zoning policies requires developers to set aside a portion of new residential developments for affordable housing units or contribute to affordable housing funds as a condition of approval. Inclusionary zoning programs leverage private development to create mixed-income communities and increase affordable housing opportunities in high-cost areas. Enacting mandatory inclusionary zoning ordinances, offering density bonuses or regulatory incentives, and providing technical assistance to developers can encourage the production of affordable housing units and promote socioeconomic diversity in neighborhoods.

Community Land Trusts (CLTs):

Establishing community land trusts (CLTs) enables communities to acquire and steward land for permanently affordable housing and community development purposes. CLTs maintain ownership of land while selling or leasing housing units at affordable prices to income-qualified buyers or renters. Supporting the expansion of CLTs, providing funding for land acquisition and infrastructure development, and promoting resident participation in CLT governance can preserve affordable housing stock, prevent displacement, and empower communities to control their own housing destinies.

Tenant Protections and Rent Control: Enacting tenant protections and rent control measures safeguards tenants from unjust eviction, excessive rent increases, and housing instability. Tenant protection laws, such as just-cause eviction policies, rent stabilization ordinances, and anti-retaliation provisions, provide tenants with legal rights and remedies to address housing-related grievances and maintain stable housing arrangements. Strengthening tenant protections, enforcing housing codes, and providing access to legal assistance and tenant advocacy services can prevent housing discrimination and ensure housing stability for renters.

Housing First Approach to Homelessness: Adopting a Housing First approach to homelessness prioritizes providing permanent supportive housing to individuals experiencing homelessness without preconditions or barriers to entry.

134

Housing First programs offer stable housing and wraparound support services, such as case management, mental health treatment, and substance abuse counseling, to help individuals maintain housing stability and achieve long-term self-sufficiency. Investing in Housing First initiatives, expanding supportive housing models, and coordinating cross-sector collaborations can effectively address chronic homelessness and promote housing stability for vulnerable populations.

Preservation of Affordable Housing Stock: Preserving existing affordable housing stock is essential for maintaining housing affordability and preventing displacement in gentrifying neighborhoods. Implementing policies that protect rent-controlled units, enforce affordability covenants, and provide financial incentives for landlords to maintain affordable rents can prevent the loss of affordable housing units due to market pressures and speculative investment. Supporting nonprofit affordable housing developers, offering property tax incentives, and leveraging public-private

partnerships can facilitate the preservation of affordable housing stock and promote neighborhood stability.

By implementing affordable housing initiatives that prioritize housing affordability, accessibility, and community development, policymakers can address the root causes of the wealth gap, promote economic opportunity, and ensure housing security for all Americans. Through collaborative efforts and a commitment to equitable housing policies, we can create vibrant and inclusive communities where everyone has access to safe, stable, and affordable housing options.

Education and Workforce Development

Universal Access to Early Childhood Education: Providing universal access to high-quality early childhood education programs ensures that all children have a strong foundation for learning and development. Investing in early childhood education initiatives, such as pre-kindergarten programs, home visiting services, and early intervention programs, can improve school readiness, narrow achievement gaps, and promote long-term educational success for children from disadvantaged backgrounds.

Equitable K-12 Education Funding: Ensuring equitable funding for K-12 education promotes educational equity and reduces disparities in educational outcomes and opportunities. Reforming school funding formulas to allocate resources based on student needs, reducing reliance on property taxes for school funding, and increasing investments in high-poverty schools can improve educational quality, support teacher recruitment and retention, and enhance student achievement for underserved communities.

College Affordability and Student Debt Relief: Making college more affordable and reducing student debt burdens is essential for expanding access to higher education and promoting economic mobility. Implementing tuition-free or debt-free college programs, increasing need-based financial aid, and expanding access to income-driven repayment plans and loan forgiveness programs can alleviate financial barriers to

137

higher education and empower students to pursue postsecondary credentials and degrees.

Career and Technical Education (CTE) Programs: Expanding access to career and technical education (CTE) programs provides students with hands-on training, industry certifications, and career pathways in high-demand fields. Investing in CTE programs, establishing partnerships with employers and industry stakeholders, and offering work-based learning opportunities can prepare students for successful transitions to the workforce, meet labor market demands, and promote economic prosperity in local communities.

Workforce Development Initiatives: Implementing workforce development initiatives provides individuals with the skills, training, and credentials needed to succeed in the 21st-century economy. Offering job training programs, apprenticeships, and vocational rehabilitation services, particularly for underserved populations and individuals with barriers to employment, can increase workforce participation, reduce

unemployment, and address skills gaps in key industries and sectors.

Support for Lifelong Learning and Upskilling: Promoting lifelong learning and upskilling opportunities enables individuals to adapt to changing labor market demands and pursue continuous professional development throughout their careers. Investing in adult education programs, online learning platforms, and workforce training vouchers can empower workers to acquire new skills, explore career pathways, and remain competitive in the evolving job market.

Promotion of STEM Education and Innovation: Fostering STEM (science, technology, engineering, and mathematics) education and innovation prepares students for careers in high-growth industries and drives economic growth and competitiveness. Expanding access to STEM education programs, supporting STEM curriculum development, and investing in STEM research and innovation can cultivate a skilled workforce, spur technological advancements, and promote economic innovation and prosperity.

139

Inclusive Education and Diversity Initiatives: Promoting inclusive education and diversity initiatives ensures that all students have access to equitable educational opportunities and resources. Implementing culturally responsive teaching practices, fostering inclusive learning environments, and promoting diversity in educational leadership and faculty recruitment can enhance educational outcomes, promote social justice, and prepare students to thrive in a diverse and interconnected world.

Entrepreneurship and Small Business Development: Supporting entrepreneurship and small business development fosters economic growth, job creation, and innovation in local communities. Providing access to entrepreneurship training, business development resources, and small business loans and grants can empower aspiring entrepreneurs, particularly women, minorities, and disadvantaged individuals, to start and grow successful businesses, build wealth, and contribute to economic revitalization efforts.

140

Financial Literacy and Career Counseling:

Promoting financial literacy and career counseling services equips individuals with the knowledge, skills, and resources needed to make informed decisions about education, careers, and financial planning. Offering financial education workshops, one-on-one counseling sessions, and online resources can improve financial literacy, empower individuals to navigate educational and career pathways, and promote economic self-sufficiency and well-being.

By prioritizing education and workforce development initiatives that promote equity, access, and opportunity for all individuals, policymakers, educators, and community leaders can address the root causes of the wealth gap and create a more inclusive and equitable economy where everyone has the opportunity to thrive and succeed. Through collaborative efforts and a commitment to lifelong learning, we can build a brighter future for generations to come.

Access to Quality Education

Equitable Funding: Ensuring equitable funding for schools is essential for providing all students with access to quality education. Reforming school funding formulas to allocate resources based on student needs, reducing reliance on property taxes for school funding, and increasing investments in high-poverty schools can improve educational quality and promote equitable outcomes for students from diverse socioeconomic backgrounds.

Universal Pre-Kindergarten:

Implementing universal pre-kindergarten programs ensures that all children have access to early childhood education opportunities that promote school readiness and academic success. Investing in high-quality pre-kindergarten programs, expanding access to early childhood education initiatives, and providing support services for children and families can narrow achievement gaps and foster positive developmental outcomes for children from disadvantaged backgrounds.

Reducing Class Sizes: Reducing

class sizes allows teachers to provide individualized attention and support to students, promote student engagement and participation, and address diverse learning needs. Investing in smaller class sizes, hiring additional teachers and support staff, and implementing effective instructional strategies can improve academic performance, enhance student outcomes, and create more supportive learning environments for all students.

High-Quality Teachers and Professional Development:

Recruiting and retaining high-quality teachers and providing ongoing professional development opportunities are critical for improving educational outcomes and promoting student success. Offering competitive salaries, providing mentoring and coaching programs, and supporting teacher collaboration and professional learning communities can enhance teacher effectiveness, promote instructional excellence, and strengthen school communities.

Curriculum and Instructional Resources: Providing access to high-quality curriculum materials, instructional resources, and technology tools supports effective teaching and learning practices in classrooms. Investing in curriculum development, updating instructional

143

materials, and expanding access to digital learning resources can enhance student engagement, foster critical thinking skills, and prepare students for success in college and careers in the 21st-century economy.

Support for Special Education and English Language Learners (ELL): Ensuring support for special education students and English language learners (ELL) promotes educational equity and inclusion. Providing specialized instruction, accommodations, and support services for students with disabilities and English language learners, as well as offering professional development opportunities for educators, can address diverse learning needs and promote academic achievement for all students.

Social and Emotional Learning (SEL): Integrating social and emotional learning (SEL) into school curricula promotes positive youth development, enhances interpersonal skills, and fosters a positive school climate. Implementing SEL programs, providing training for educators, and creating supportive school

144

environments that prioritize emotional well-being and relationship-building can improve student behavior, reduce disciplinary incidents, and enhance academic engagement and achievement.

Access to Advanced Placement (AP) and Dual Enrollment Programs:

Expanding access to Advanced Placement (AP) courses, dual enrollment programs, and college-level coursework enables students to earn college credit and accelerate their academic progress. Providing financial support for AP exams, offering dual enrollment partnerships with colleges and universities, and ensuring equitable access to advanced coursework can increase college readiness, promote academic rigor, and broaden opportunities for students to pursue higher education and career pathways.

Closing the Digital Divide:

Closing the digital divide ensures that all students have access to technology tools, internet connectivity, and digital resources that support learning both in and out of the classroom. Investing in technology infrastructure, providing devices and broadband access to students and families, and offering digital literacy training programs can bridge the digital divide, promote digital equity, and enhance educational opportunities for underserved communities.

Parent and Community Engagement:
Engaging parents and communities as partners in education promotes collaborative relationships, strengthens school-home connections, and supports student success. Establishing parent involvement initiatives, creating opportunities for community engagement and collaboration, and fostering inclusive decision-making processes can build trust, enhance communication, and create supportive school environments that empower students to reach their full potential.

By prioritizing access to quality education for all students, policymakers, educators, and community stakeholders can address the root causes of the wealth gap and create a more equitable and inclusive society where every individual has the opportunity to thrive and succeed. Through collective efforts and a commitment to educational equity, we can build a brighter future for generations to come.

Job Training and Skill Development Programs

Workforce Development Initiatives: Implementing workforce development initiatives provides individuals with the skills, training, and credentials needed to succeed in the 21st-century economy. These initiatives include job training programs, apprenticeships, and vocational rehabilitation services designed to equip participants with industry-specific skills and certifications that align with employer needs and labor market demands.

Apprenticeship Programs: Establishing apprenticeship programs offers hands-on training and mentorship opportunities for individuals seeking to enter skilled trades and technical occupations. Apprenticeships combine classroom instruction with paid on-the-job training, allowing participants to earn while they learn and acquire industry-recognized credentials that enhance their employability and career advancement prospects.

147

Career Pathways and Sector-Based Training: Promoting career pathways and sector-based training models provides individuals with clear pathways to high-demand industries and occupations. These programs offer structured training programs, career counseling services, and support services such as childcare and transportation assistance to help participants overcome barriers to employment and succeed in targeted industry sectors.

Adult Education and Literacy Programs: Investing in adult education and literacy programs enables individuals to improve their basic skills, earn high school equivalency credentials, and pursue further education and training opportunities. Adult education programs offer flexible learning options, personalized instruction, and wraparound support services to help adults build foundational skills, transition to postsecondary education or training, and access career pathways with family-sustaining wages.

Digital Skills and Technology Training: Providing digital skills and

technology training programs equips individuals with the technical skills and digital literacy needed to thrive in an increasingly digital economy. These programs offer instruction in areas such as computer programming, software development, digital marketing, and cybersecurity, as well as opportunities to earn industry-recognized certifications and credentials that enhance employability and career advancement opportunities.

Entrepreneurship and Small Business Development: Supporting entrepreneurship and small business development initiatives empowers individuals to start and grow successful businesses, create jobs, and contribute to local economic development. Entrepreneurship programs offer training, mentoring, and access to resources such as business planning assistance, financing options, and networking opportunities that help aspiring entrepreneurs navigate the complexities of starting and managing a business.

Financial Literacy and Workforce Readiness Training: Promoting financial literacy and workforce

readiness training equips individuals with essential skills and knowledge needed to make informed decisions about education, careers, and financial planning. These programs offer instruction in areas such as budgeting, savings, credit management, job search strategies, resume writing, and interview skills, as well as access to resources and support services that promote economic self-sufficiency and financial well-being.

Reentry and Second-Chance Programs:
Providing reentry and second-chance programs offers opportunities for individuals with criminal records to access education, training, and employment services that support successful reintegration into society. These programs offer wraparound support services, job readiness training, and transitional employment opportunities that help individuals overcome barriers to employment, build positive relationships, and achieve long-term stability and self-sufficiency.

Work-Based Learning and Internship Programs:
Implementing work-based learning and internship

150

programs provides students and job seekers with real-world work experience, industry connections, and practical skills that enhance their employability and career readiness. These programs offer hands-on learning opportunities, mentorship, and exposure to workplace environments that help participants develop professional skills, build professional networks, and explore career pathways aligned with their interests and aspirations.

Targeted Support for Disadvantaged Populations:

Targeting support for disadvantaged populations, including low-income individuals, minorities, veterans, individuals with disabilities, and displaced workers, ensures that all individuals have access to opportunities for skill development and career advancement. These programs offer tailored services, specialized supports, and outreach efforts that address the unique needs and barriers faced by marginalized populations and promote equitable access to economic opportunity and upward mobility.

151

By prioritizing job training and skill development programs that promote equity, access, and opportunity for all individuals, policymakers, educators, and workforce development practitioners can address the root causes of the wealth gap and create a more inclusive and equitable economy where everyone has the opportunity to thrive and succeed. Through collaborative efforts and a commitment to lifelong learning, we can build a brighter future for generations to come.

Promoting Entrepreneurship and Small Business Ownership

Access to Capital: Improving access to capital for aspiring entrepreneurs and small business owners is crucial for promoting entrepreneurship and fostering business growth. Initiatives such as microloans, small business grants, and alternative financing options can provide startup capital and working capital for entrepreneurs, particularly those from underserved communities and minority-owned businesses.

152

Entrepreneurship Education and Training: Providing entrepreneurship education and training programs equips aspiring entrepreneurs with the knowledge, skills, and resources needed to start and grow successful businesses. Entrepreneurship programs offer instruction in areas such as business planning, marketing, financial management, and legal compliance, as well as access to mentorship, networking opportunities, and business incubation services that support business development and sustainability.

Technical Assistance and Business Support Services: Offering technical assistance and business support services provides entrepreneurs with personalized guidance, mentorship, and expertise to navigate the complexities of starting and managing a business. Business assistance programs offer one-on-one counseling, advisory services, and access to professional resources such as legal, accounting, and marketing support that help entrepreneurs overcome challenges, make informed decisions, and achieve business success.

153

Access to Markets and Procurement Opportunities:

Facilitating access to markets and procurement opportunities enables small businesses to compete for contracts, partnerships, and opportunities in the public and private sectors. Supplier diversity initiatives, government contracting programs, and matchmaking events connect small businesses with potential buyers, vendors, and strategic partners, while promoting economic inclusion and diversity in procurement practices.

Minority and Women-Owned Business Development:

Supporting minority-owned and women-owned businesses promotes economic empowerment, diversity, and inclusion in the entrepreneurial ecosystem. Minority business development centers, women's business centers, and minority supplier development programs offer targeted resources, training, and networking opportunities that help minority and women entrepreneurs overcome barriers to success and thrive in competitive markets.

154

Incubators and Accelerators:
Establishing incubators and accelerators provides startups and early-stage ventures with resources, mentorship, and support services that accelerate growth and innovation. Incubator and accelerator programs offer access to shared office space, networking events, investor pitch opportunities, and industry expertise that help entrepreneurs validate ideas, develop products, and scale their businesses more quickly and effectively.

Regulatory Reform and Simplification:
Streamlining regulatory processes and reducing bureaucratic barriers simplifies the business environment and encourages entrepreneurship and small business development. Regulatory reform initiatives, such as business licensing reform, permit simplification, and compliance assistance programs, help entrepreneurs navigate regulatory requirements, reduce administrative burdens, and focus on building and growing their businesses.

Technology and Innovation Ecosystems:
Cultivating technology

and innovation ecosystems fosters entrepreneurship, collaboration, and economic growth in emerging industries and sectors. Innovation hubs, technology parks, and research institutions provide entrepreneurs with access to research and development resources, technology infrastructure, and startup support services that facilitate innovation, attract investment, and drive economic development in local communities.

Community Development and Economic Empowerment: Investing in community development and economic empowerment initiatives strengthens local economies, creates jobs, and promotes entrepreneurship in underserved communities. Community development financial institutions (CDFIs), community development corporations (CDCs), and economic empowerment zones offer financial products, technical assistance, and capacity-building support that empower entrepreneurs to launch businesses, create wealth, and revitalize neighborhoods.

Ecosystem Collaboration and Partnership: Fostering collaboration

156

and partnership among stakeholders in the entrepreneurial ecosystem promotes collective action, resource sharing, and innovation-driven economic development. Building strategic partnerships among government agencies, educational institutions, industry associations, and community organizations fosters an ecosystem of support that nurtures entrepreneurship, fosters business growth, and creates pathways to prosperity for all stakeholders.

By promoting entrepreneurship and small business ownership through targeted policies, programs, and initiatives, policymakers, business leaders, and community stakeholders can empower individuals to pursue economic independence, create wealth, and contribute to inclusive economic growth. Through collaborative efforts and a commitment to fostering an entrepreneurial culture of innovation and opportunity, we can address the wealth gap and create a more equitable and prosperous economy for future generations.

Community and Grassroots Initiatives

Community Wealth Building:

Community wealth building initiatives empower residents to collectively own and control local assets, businesses, and resources that generate wealth and promote economic self-sufficiency. Community land trusts, worker cooperatives, and community development corporations (CDCs) foster community ownership, democratic decision-making, and shared prosperity by reinvesting profits and resources back into the community to address social and economic challenges.

Local Investment Networks:

Establishing local investment networks enables residents to pool resources, capital, and expertise to finance community-based projects and ventures that support economic development and revitalization. Local investment clubs, community loan funds, and impact investing networks provide alternative financing options and investment opportunities that promote economic resilience, entrepreneurship, and job creation in underserved communities.

Community Development Partnerships:
Building collaborative partnerships among stakeholders fosters collective action, resource sharing, and innovation-driven solutions to address community needs and priorities. Community development partnerships bring together government agencies, nonprofits, businesses, and residents to identify assets, leverage resources, and implement coordinated strategies that promote equitable development, social inclusion, and community well-being.

Neighborhood Revitalization Initiatives:
Implementing neighborhood revitalization initiatives revitalizes distressed communities, strengthens local economies, and improves quality of life for residents. Neighborhood revitalization efforts focus on affordable housing development, commercial corridor revitalization, infrastructure improvements, and placemaking activities that enhance neighborhood aesthetics, attract investment, and create vibrant and

inclusive communities where residents can live, work, and thrive.

Community-Based Entrepreneurship Programs:

Supporting community-based entrepreneurship programs empowers residents to start and grow businesses that meet local needs, create jobs, and generate wealth within the community. Entrepreneurship programs offer business training, technical assistance, and access to capital that help aspiring entrepreneurs overcome barriers to success and launch businesses that contribute to economic development and community prosperity.

Place-Based Initiatives:

Implementing place-based initiatives targets investments and resources to specific geographic areas with concentrated poverty and disinvestment. Place-based strategies focus on comprehensive community development, neighborhood stabilization, and asset-based approaches that leverage local assets, strengths, and cultural resources to promote economic opportunity, social equity, and resident engagement in distressed neighborhoods and communities.

Community Land Trusts (CLTs):

Establishing community land trusts (CLTs) enables communities to acquire and steward land for affordable housing, commercial development, and community amenities that serve the needs of residents and promote long-term affordability and stability. CLTs empower residents to collectively own and control land, preserve affordability, and prevent displacement in rapidly gentrifying neighborhoods while fostering community ownership and empowerment.

Social Enterprise Development:

Promoting social enterprise development encourages the growth of businesses and organizations that prioritize social impact, community benefit, and environmental sustainability alongside financial profitability. Social enterprises reinvest profits into mission-driven activities, such as workforce development, environmental conservation, and social services, that address pressing social and economic challenges and create shared value for communities and stakeholders.

Community-Led Planning and Decision-Making: Engaging residents in community-led planning and decision-making processes promotes participatory democracy, civic engagement, and resident empowerment. Community-led planning initiatives involve residents in identifying priorities, setting goals, and shaping strategies for inclusive and equitable development that reflect the needs, aspirations, and values of diverse community members.

Equitable Development Policies: Adopting equitable development policies ensures that economic growth and investment benefit all residents and communities, particularly those historically marginalized or underserved. Equitable development policies prioritize affordable housing, inclusive economic development, and anti-displacement measures that promote social equity, racial justice, and shared prosperity for all residents, regardless of income or background.

By promoting community and grassroots initiatives that prioritize equity, inclusion, and resident empowerment, policymakers, community leaders, and grassroots

162

organizers can address the root causes of the wealth gap and build a more equitable and resilient economy that benefits everyone. Through collective action and a commitment to community-driven solutions, we can create vibrant, thriving, and inclusive communities where everyone has the opportunity to succeed and prosper.

Community Investment Funds

Introduction: Community investment funds represent a powerful tool for addressing the wealth gap and fostering economic equity in the United States. These funds pool capital from various sources and direct investments toward projects, businesses, and initiatives that promote community development, create jobs, and generate wealth in underserved communities. By leveraging financial resources and local expertise, community investment funds empower residents to drive positive change, build assets, and unlock economic opportunities in their neighborhoods.

Key Components and Characteristics:

Diverse Capital Sources:
Community investment funds source
capital from a variety of stakeholders,
including individual investors, philanthropic
organizations, government agencies,
financial institutions, and community
development financial institutions (CDFIs).
By tapping into diverse funding streams,
these funds mobilize resources to support
community-driven initiatives and projects
that align with social and environmental
objectives.

Mission-Driven Investing:
Community investment funds prioritize
mission-driven investing that generates
positive social, environmental, and
economic impacts alongside financial
returns. These funds adhere to socially
responsible investment criteria and
prioritize investments in affordable
housing, small businesses, renewable
energy, sustainable agriculture, and other
sectors that advance community
development goals and address systemic
inequities.

**Local Decision-Making and
Governance:** Community investment
funds operate with a focus on local

164

decision-making and governance structures that ensure accountability, transparency, and community representation. Boards of directors, investment committees, and advisory councils typically comprise local stakeholders, residents, and experts who oversee fund activities, assess investment opportunities, and align fund strategies with community needs and priorities.

Targeted Investment Strategies:
Community investment funds deploy targeted investment strategies that address specific challenges and opportunities in underserved communities. These strategies may include affordable housing development, small business lending, workforce training, infrastructure financing, and neighborhood revitalization initiatives that catalyze economic growth, promote social equity, and enhance quality of life for residents.

Risk Management and Impact Measurement: Community investment funds employ rigorous risk management and impact measurement practices to assess the social, environmental, and financial performance of investments.

165

Through impact evaluation frameworks, outcome metrics, and reporting mechanisms, these funds track progress, evaluate effectiveness, and demonstrate accountability to investors, stakeholders, and the communities they serve.

Partnerships and Collaborations:
Community investment funds forge strategic partnerships and collaborations with local organizations, businesses, government agencies, and investors to leverage resources, share expertise, and amplify impact. These partnerships facilitate knowledge exchange, co-investment opportunities, and collective action that mobilize capital, scale initiatives, and address complex challenges facing underserved communities.

Inclusive and Equitable Investment Criteria:
Community investment funds incorporate inclusive and equitable investment criteria that prioritize projects and businesses owned or led by women, minorities, veterans, immigrants, and other historically marginalized groups. By promoting diversity, equity, and inclusion in investment decision-making

166

processes, these funds foster economic empowerment, reduce disparities, and build pathways to prosperity for underrepresented entrepreneurs and communities.

Long-Term Commitment to Community Development:

Community investment funds demonstrate a long-term commitment to community development and economic resilience by deploying patient capital, fostering sustainable business models, and supporting capacity-building efforts that strengthen local institutions and ecosystems. Through continuous engagement, adaptive strategies, and responsive investments, these funds catalyze transformative change and create lasting impact in underserved communities.

Case Studies and Success Stories:

The Calvert Foundation's Community Investment Note: This innovative investment vehicle allows individual investors to channel capital into community development projects and initiatives across the United States,

167

including affordable housing, small business lending, and renewable energy projects.

The Opportunity Finance Network (OFN): OFN serves as a national network of CDFIs and community development lenders that deploy capital to support low-income and underserved communities. OFN's member institutions provide financing, technical assistance, and capacity-building support to entrepreneurs, nonprofits, and affordable housing developers.

The New Markets Tax Credit (NMTC) Program: This federal tax incentive program incentivizes private investment in distressed communities by providing tax credits to investors who finance qualified community development projects, such as commercial real estate developments, small business expansions, and healthcare facilities.

Conclusion: Community investment funds represent a powerful mechanism for mobilizing capital, driving economic development, and advancing equity in the United States. By harnessing the collective power of investors, stakeholders, and communities, these funds create pathways to prosperity, foster inclusive growth, and build resilient economies that benefit all residents. As we strive to address the wealth gap and create a more equitable society, community investment funds offer a compelling model for driving positive change and unlocking opportunities for shared prosperity.

Microfinance and Credit Programs

Introduction: Microfinance and credit programs play a crucial role in addressing the wealth gap by providing access to financial services, capital, and resources for underserved individuals and communities in the United States. These programs empower entrepreneurs, small business owners, and low-income households to build assets, create economic opportunities, and achieve financial stability. By promoting financial inclusion, entrepreneurship, and wealth-building, microfinance and credit programs contribute to a more equitable and inclusive economy where all individuals have the opportunity to thrive and prosper.

Key Components and Characteristics:

Financial Inclusion and Access: Microfinance and credit programs promote financial inclusion by expanding access to affordable financial products and services for individuals who are underserved or excluded from traditional banking systems. These programs offer savings accounts, checking accounts, credit lines, and other financial tools that empower individuals to manage their finances, build credit histories, and access capital to pursue economic opportunities.

Microcredit and Small Business Lending:

Microfinance institutions (MFIs) and community development financial institutions (CDFIs) provide microcredit and small business loans to entrepreneurs, startups, and small business owners who lack access to conventional bank loans or financing options. Microcredit programs offer small, short-term loans with flexible terms, low-interest rates, and simplified application processes that enable borrowers to invest in their businesses, purchase equipment, and expand operations.

170

Peer Lending and Group-Based Models:

Peer lending and group-based lending models leverage social capital and community networks to facilitate access to credit for individuals who have limited collateral or credit history. These models involve forming lending groups, peer support networks, or solidarity lending circles where borrowers collectively guarantee each other's loans and provide mutual support, accountability, and encouragement to repay debts and achieve financial goals.

Financial Education and Capacity Building: Microfinance and credit programs offer financial education, literacy training, and capacity-building workshops that equip borrowers with the knowledge, skills, and resources needed to make informed financial decisions, manage household budgets, and navigate the financial system effectively. Financial education programs cover topics such as budgeting, savings, debt management, credit building, and entrepreneurship to empower individuals to achieve financial independence and economic self-sufficiency.

171

Asset-Building and Wealth Creation: Microfinance and credit programs promote asset-building and wealth creation by enabling individuals to accumulate savings, build creditworthiness, and access capital to invest in income-generating assets, such as homes, businesses, education, and healthcare. These programs facilitate wealth accumulation and intergenerational transfers of assets that contribute to long-term financial stability, economic mobility, and prosperity for individuals and families.

Targeted Outreach and Support Services: Microfinance and credit programs engage in targeted outreach and provide support services to reach underserved populations, including low-income individuals, women, minorities, immigrants, veterans, and individuals with disabilities. Outreach efforts involve partnering with community organizations, faith-based institutions, and local stakeholders to raise awareness, build trust, and address barriers to access that hinder financial inclusion and participation in mainstream financial markets.

172

Social Impact Investing and Responsible Finance: Microfinance and credit programs embrace social impact investing principles and responsible finance practices that prioritize positive social, environmental, and economic outcomes alongside financial returns. These programs adhere to ethical lending standards, fair lending practices, and transparent governance structures that ensure accountability, sustainability, and alignment with community development goals and values.

Data Analytics and Technology Innovation: Microfinance and credit programs leverage data analytics, technology innovation, and digital platforms to streamline operations, enhance customer experiences, and expand reach and scale. Digital banking solutions, mobile payment platforms, and online lending portals enable borrowers to access financial services remotely, conduct transactions securely, and manage accounts conveniently, thereby reducing barriers to access and improving financial inclusion outcomes.

Case Studies and Success Stories:

173

Grameen America: Grameen America is a leading microfinance organization that provides microloans, financial education, and support services to low-income women entrepreneurs across the United States. By leveraging the Grameen Bank's group-based lending model, Grameen America has empowered thousands of women to start and expand businesses, increase household incomes, and achieve economic independence.

Accion Opportunity Fund: Accion Opportunity Fund is a nonprofit CDFI that offers small business loans, technical assistance, and financial coaching to underserved entrepreneurs and communities in the United States. Through its lending programs and business support services, Accion Opportunity Fund has helped entrepreneurs access capital, create jobs, and stimulate economic growth in low-income neighborhoods and minority-owned businesses.

Kiva US: Kiva US is a nonprofit crowdfunding platform that connects lenders with borrowers to facilitate peer-to-peer microloans for small business owners, artisans, and individuals with

174

entrepreneurial aspirations. By harnessing the power of collective giving and social capital, Kiva US has provided thousands of borrowers with access to affordable capital, mentorship, and community support to pursue their dreams and build sustainable businesses.

Conclusion: Microfinance and credit programs play a vital role in fostering economic inclusion, promoting entrepreneurship, and reducing the wealth gap in the United States. By expanding access to financial services, empowering underserved individuals and communities, and promoting responsible finance practices, these programs contribute to a more equitable and inclusive economy where everyone has the opportunity to achieve financial security, pursue their aspirations, and build a brighter future for themselves and their families. Through collaborative efforts and a commitment to financial inclusion and empowerment, we can create a more just and prosperous society where economic opportunity is accessible to all.

Local Economic Development Projects

Introduction: Local economic development projects play a crucial role in addressing the wealth gap by stimulating economic growth, creating jobs, and fostering community prosperity in underserved neighborhoods and communities across the United States. These projects leverage local assets, resources, and partnerships to revitalize distressed areas, expand economic opportunities, and improve quality of life for residents. By promoting inclusive and sustainable development strategies, local economic development projects contribute to building a more equitable and resilient economy where everyone has the opportunity to thrive and succeed.

Key Components and Characteristics:

Strategic Planning and Visioning: Local economic development projects begin with strategic planning and visioning processes that engage stakeholders, residents, and community leaders in defining priorities, setting goals, and shaping strategies for

176

inclusive and sustainable growth. These processes involve conducting community assessments, identifying assets and opportunities, and developing action plans that align with community needs and aspirations.

Infrastructure Investment and Redevelopment:

Local economic development projects invest in infrastructure improvements and redevelopment initiatives that enhance the physical environment, attract private investment, and catalyze economic activity in targeted areas. Infrastructure projects may include transportation enhancements, utility upgrades, streetscape improvements, and brownfield remediation efforts that create a foundation for business expansion, job creation, and community revitalization.

Business Attraction and Retention:

Local economic development projects focus on attracting and retaining businesses that contribute to job creation, innovation, and economic diversity in the community. Economic development agencies, business improvement districts, and chambers of commerce engage in

177

marketing, business recruitment, and retention efforts to identify prospective employers, support existing businesses, and facilitate expansion opportunities that strengthen the local economy and generate tax revenue.

Entrepreneurship and Small Business Support:
Local economic development projects support entrepreneurship and small business development through access to capital, technical assistance, and support services that help aspiring entrepreneurs launch and grow successful ventures. Business incubators, accelerators, and entrepreneurship centers provide mentoring, training, and networking opportunities that foster innovation, encourage startup formation, and cultivate a culture of entrepreneurship within the community.

Workforce Development and Training:
Local economic development projects invest in workforce development and training programs that equip residents with the skills, knowledge, and credentials needed to compete in the 21st-century economy. Workforce development

initiatives offer career counseling, job placement services, and vocational training programs that align with industry needs, promote career advancement, and address skill gaps in high-demand sectors.

Community-Based Tourism and Cultural Development:

Local economic development projects leverage tourism and cultural assets to promote community identity, attract visitors, and generate economic opportunities for local businesses and artisans. Cultural districts, heritage trails, and arts festivals showcase the unique history, culture, and creativity of the community, while supporting small businesses, artists, and cultural entrepreneurs that contribute to the local economy and enrich the quality of life for residents.

Green and Sustainable Development Practices:

Local economic development projects embrace green and sustainable development practices that promote environmental stewardship, resource conservation, and climate resilience while fostering economic growth and social equity. Sustainable development initiatives

179

prioritize energy efficiency, renewable energy adoption, green infrastructure, and environmentally-friendly practices that create green jobs, reduce carbon emissions, and enhance community sustainability and livability.

Affordable Housing and Neighborhood Revitalization: Local economic development projects prioritize affordable housing and neighborhood revitalization efforts that promote housing affordability, combat displacement, and create inclusive communities where residents of all income levels can live, work, and thrive. Affordable housing developments, mixed-use projects, and community land trusts preserve affordability, promote housing stability, and contribute to vibrant, diverse neighborhoods that support economic inclusion and social cohesion.

Public-Private Partnerships and Collaboration: Local economic development projects foster public-private partnerships and collaboration among stakeholders, government agencies, nonprofit organizations, and private sector investors to leverage resources, share expertise, and advance shared goals and priorities. Collaborative approaches to

180

economic development enable cross-sectoral coordination, maximize impact, and address complex challenges that require collective action and community engagement.

Equity, Inclusion, and Social Justice:
Local economic development projects prioritize equity, inclusion, and social justice principles that ensure that economic opportunities and benefits are distributed equitably and accessible to all residents, particularly those from historically marginalized or underserved communities. These projects employ inclusive planning processes, equitable investment strategies, and anti-displacement measures that promote fairness, diversity, and social equity in economic development outcomes and decision-making processes.

Case Studies and Success Stories:

The Riverwalk Project in San Antonio, Texas:
This transformative urban revitalization project transformed a neglected stretch of the San Antonio River into a vibrant cultural and recreational destination that attracts millions of visitors annually, supports local businesses, and

181

generates economic activity in downtown San Antonio.

The Brooklyn Navy Yard Development Corporation in New York City:

The Brooklyn Navy Yard Development Corporation revitalized a historic naval shipyard into a modern industrial park and innovation hub that provides affordable space for manufacturing, technology, and creative industries, creating thousands of jobs and fostering entrepreneurship and innovation in Brooklyn's waterfront communities.

The West Louisville FoodPort in Louisville, Kentucky: The West Louisville FoodPort initiative aims to revitalize a vacant brownfield site into a sustainable urban food hub that supports local agriculture, promotes healthy food access, and creates economic opportunities for residents in Louisville's underserved neighborhoods.

Conclusion:

Local economic development projects are critical drivers of inclusive and sustainable growth that promote economic opportunity, community prosperity, and social equity in cities and regions across the United States. By investing in infrastructure,

182

supporting entrepreneurship, and fostering collaboration, these projects empower residents, businesses, and stakeholders to shape the future of their communities and build a more equitable and resilient economy for all. Through visionary leadership, strategic partnerships, and community engagement, we can address the wealth gap and create thriving, vibrant, and inclusive communities where everyone has the opportunity to succeed and prosper.

Corporate Responsibility and Accountability

Introduction: Corporate responsibility and accountability are essential components in addressing the wealth gap and fostering a more equitable economy in the United States. As major drivers of economic activity, corporations wield significant influence over social, environmental, and economic outcomes. By embracing principles of corporate responsibility and accountability, businesses can play a pivotal role in promoting inclusive growth, reducing inequality, and advancing social justice. This chapter explores the importance of

183

corporate responsibility, key strategies for enhancing accountability, and the potential impact of corporate engagement on addressing systemic inequities and creating shared prosperity.

Key Components and Characteristics:

Stakeholder Engagement and Dialogue: Corporate responsibility begins with meaningful engagement and dialogue with stakeholders, including employees, customers, communities, investors, and civil society organizations. By listening to diverse perspectives and understanding stakeholder needs and concerns, corporations can identify opportunities to align business practices with societal expectations, values, and priorities.

Ethical Leadership and Governance: Ethical leadership and governance are fundamental to corporate responsibility and accountability. Corporate leaders set the tone for organizational culture, values, and behavior by demonstrating integrity, transparency, and a commitment to ethical business practices. Strong governance structures, independent oversight mechanisms, and

184

accountability frameworks help ensure that corporate decisions and actions align with ethical standards and legal requirements.

Environmental Sustainability and Stewardship: Corporate

responsibility encompasses environmental sustainability and stewardship practices that minimize ecological impact, conserve natural resources, and mitigate climate change. Sustainable business strategies, energy efficiency measures, waste reduction initiatives, and carbon footprint reduction efforts enable corporations to operate more responsibly and contribute to a healthier, more sustainable planet for future generations.

Social Impact and Community Investment: Corporate responsibility

includes social impact and community investment initiatives that address pressing social issues, support community development, and improve quality of life for individuals and families. Corporate philanthropy, volunteer programs, and community partnerships enable businesses to make meaningful contributions to education, healthcare, affordable housing, workforce development, and other areas of critical need.

Diversity, Equity, and Inclusion (DEI) Practices:

Corporate responsibility encompasses diversity, equity, and inclusion (DEI) practices that foster a culture of belonging, respect, and fairness within the workplace and beyond. Diversity initiatives, inclusive hiring practices, pay equity measures, and employee resource groups promote diversity of thought, representation, and opportunity, while combating discrimination, bias, and systemic inequities in the workplace and society.

Labor Rights and Workplace Conditions:

Corporate responsibility includes upholding labor rights and ensuring fair and safe working conditions for employees throughout the supply chain. Respect for human rights, adherence to labor standards, and compliance with international labor conventions protect workers from exploitation, discrimination, and unsafe working conditions, while promoting dignity, empowerment, and economic security for all employees.

186

Transparency and Accountability Reporting:

Corporate responsibility entails transparency and accountability reporting practices that provide stakeholders with clear, accurate, and timely information about corporate performance, impact, and governance practices. Sustainability reports, corporate social responsibility (CSR) disclosures, and impact assessments enable stakeholders to assess corporate contributions to social and environmental goals, track progress over time, and hold companies accountable for their commitments and actions.

Responsible Supply Chain Management:

Corporate responsibility extends to responsible supply chain management practices that promote ethical sourcing, supplier diversity, and supply chain transparency. Supply chain due diligence, supplier audits, and supplier engagement programs help identify and address human rights violations, labor abuses, and environmental risks in global supply chains, while promoting responsible sourcing practices that benefit workers, communities, and the environment.

Corporate Advocacy and Public Policy Engagement:

Corporate responsibility includes advocacy and public policy engagement efforts that advance social and environmental goals, promote inclusive economic policies, and address systemic inequities through legislative and regulatory reform. Corporate advocacy initiatives, industry coalitions, and policy campaigns leverage corporate influence and resources to advocate for laws and policies that support human rights, environmental protection, and social justice.

Continuous Improvement and Learning:

Corporate responsibility is a journey of continuous improvement and learning that requires ongoing assessment, reflection, and adaptation to evolving social, environmental, and economic challenges. By embracing a growth mindset, fostering innovation, and embracing stakeholder feedback, corporations can drive positive change, build trust, and create long-term value for shareholders, stakeholders, and society.

Case Studies and Success Stories:

188

Patagonia: Patagonia is a leading outdoor apparel company known for its commitment to environmental sustainability, social responsibility, and corporate activism. Through initiatives such as the Patagonia Action Works platform, the company engages customers, employees, and grassroots organizations in environmental advocacy and activism campaigns that support conservation, climate action, and environmental justice.

Microsoft: Microsoft is a technology company that prioritizes diversity, equity, and inclusion (DEI) as core values within its corporate culture and operations. Through programs such as the Microsoft Supplier Diversity Program and the Diversity and Inclusion Initiative, the company promotes diversity in its workforce, supply chain, and communities, while fostering a culture of inclusion, belonging, and empowerment for all employees.

Starbucks: Starbucks is a global coffeehouse chain that integrates social impact and community engagement into its business model through initiatives such as the Starbucks College Achievement Plan,

189

which provides eligible employees with full tuition coverage for online bachelor's degree programs at Arizona State University. Starbucks also invests in sustainable coffee sourcing practices, ethical labor standards, and community development projects that empower coffee farmers and support coffee-growing communities around the world.

Conclusion:

Corporate responsibility and accountability are essential drivers of positive social change, economic inclusion, and environmental sustainability in the United States. By embracing principles of ethical leadership, sustainability, diversity, and inclusion, corporations can leverage their resources, influence, and expertise to address the wealth gap, advance equity, and create shared prosperity for all stakeholders. Through collaborative efforts and a commitment to responsible business practices, corporations can play a transformative role in shaping a more equitable and sustainable economy where everyone has the opportunity to thrive and succeed.

Fair Wages and Benefits

Introduction: Fair wages and benefits are foundational elements of a more equitable economy in the United States. Ensuring that workers receive fair compensation for their labor and have access to essential benefits not only addresses the wealth gap but also promotes economic stability, social mobility, and overall well-being. This chapter explores the importance of fair wages and benefits, challenges faced by workers, and strategies for promoting greater economic equity through fair labor practices.

Key Components and Characteristics:

Living Wage Standards: Living wage standards establish minimum compensation levels that enable workers to meet their basic needs, including housing, food, healthcare, transportation, and other essential expenses. Living wage calculations consider regional cost-of-living variations and family size to ensure that workers can afford a decent standard of living and avoid poverty despite differences in geographic location and household composition.

Minimum Wage Legislation: Minimum wage legislation sets a baseline hourly wage that employers must pay to workers, ensuring that wages remain above poverty levels and keep pace with inflation and cost-of-living increases. Increasing the minimum wage to a living wage level helps lift workers out of poverty, reduces income inequality, and stimulates consumer spending, leading to broader economic benefits for businesses and communities.

Fair Labor Standards Act (FLSA) Compliance: The Fair Labor Standards Act (FLSA) establishes federal labor standards for minimum wage, overtime pay, child labor, and recordkeeping requirements to protect workers from exploitation and ensure fair compensation for their labor. FLSA compliance helps safeguard workers' rights, prevent wage theft, and promote fair labor practices across industries and sectors.

Equal Pay for Equal Work: Equal pay for equal work principles aim to eliminate gender and racial wage gaps by ensuring that workers receive equal

192

compensation for performing substantially similar job duties and responsibilities, regardless of gender, race, ethnicity, or other protected characteristics. Pay equity measures promote fairness, diversity, and inclusion in the workplace and contribute to closing wage gaps and advancing gender and racial equality.

Living Wage Campaigns and Advocacy:
Living wage campaigns and advocacy efforts mobilize workers, unions, community organizations, and policymakers to advocate for fair wages, benefits, and working conditions that meet the needs of workers and their families. These campaigns raise awareness, build coalitions, and mobilize public support for legislative and policy reforms that promote economic justice and shared prosperity for all workers.

Unionization and Collective Bargaining:
Unionization and collective bargaining empower workers to negotiate fair wages, benefits, and working conditions through collective action and solidarity. Labor unions advocate for workers' rights, represent their interests in negotiations with employers, and secure

contractual agreements that establish fair compensation levels, healthcare coverage, retirement benefits, and other workplace protections.

Comprehensive Benefits Packages:
Comprehensive benefits packages provide workers with access to essential benefits, including healthcare coverage, paid sick leave, paid family and medical leave, retirement savings plans, disability insurance, and other workplace benefits that support their health, financial security, and work-life balance needs. Robust benefits packages enhance employee retention, morale, and productivity while promoting workforce well-being and resilience.

Income Support Programs:
Income support programs, such as the Earned Income Tax Credit (EITC) and the Child Tax Credit (CTC), provide targeted financial assistance to low- and moderate-income workers and families, supplementing their earnings and reducing economic hardship. These programs help lift families out of poverty, reduce income inequality, and stimulate economic growth by putting more

money into the hands of workers who need it most.

Corporate Social Responsibility (CSR) Commitments: Corporate social responsibility (CSR) commitments encourage businesses to adopt fair labor practices, promote worker rights, and invest in employee well-being through fair wages, benefits, and workplace policies. CSR initiatives enhance corporate reputation, foster employee engagement, and build trust with customers, investors, and stakeholders while contributing to greater economic equity and social impact.

Workforce Development and Training Programs:

Workforce development and training programs equip workers with the skills, knowledge, and credentials needed to access higher-paying jobs, advance in their careers, and command higher wages in the labor market. These programs offer vocational training, apprenticeships, certification programs, and educational opportunities that empower workers to achieve economic mobility, pursue meaningful careers, and secure financial stability for themselves and their families.

195

Case Studies and Success Stories:

Costco Wholesale: Costco Wholesale is a retail giant known for its commitment to fair wages and benefits for its employees. The company pays its workers significantly higher wages than competitors in the retail industry, offers generous healthcare coverage, retirement benefits, and opportunities for career advancement. Costco's investment in its workforce has contributed to high employee morale, low turnover rates, and strong customer loyalty, demonstrating the business case for fair labor practices.

Ben & Jerry's: Ben & Jerry's, an iconic ice cream company, is dedicated to social and economic justice, including fair wages and benefits for its employees. The company has implemented policies such as a minimum livable wage, comprehensive healthcare coverage, and paid family leave to support the well-being of its workforce. Ben & Jerry's commitment to fair labor practices aligns with its mission-driven approach to business and reflects its values of equity, sustainability, and social responsibility.

SEIU Healthcare: The Service Employees International Union (SEIU) Healthcare is a union representing healthcare workers across the United States. SEIU Healthcare advocates for fair wages, benefits, and working conditions for frontline workers in the healthcare industry, including nurses, aides, and support staff. Through collective bargaining, organizing campaigns, and advocacy efforts, SEIU Healthcare has secured better wages, healthcare coverage, and workplace protections for its members, improving the quality of jobs and care in the healthcare sector.

Conclusion:

Fair wages and benefits are essential components of a more equitable economy that promotes dignity, fairness, and opportunity for all workers. By advocating for living wages, fair labor standards, and comprehensive benefits packages, policymakers, businesses, labor unions, and advocacy organizations can address the wealth gap, reduce income inequality, and build a more just and inclusive society where everyone has the opportunity to thrive and prosper. Through collaborative

197

efforts and a commitment to fair labor practices, we can create workplaces that value and respect the contributions of workers, support their well-being, and promote shared prosperity for generations to come.

Diversity and Inclusion Initiatives

Introduction: Diversity and inclusion initiatives are integral components of efforts to address the wealth gap and foster a more equitable economy in the United States. Embracing diversity and promoting inclusion within workplaces, communities, and society at large not only promotes social justice but also drives innovation, enhances productivity, and creates opportunities for economic advancement. This chapter explores the importance of diversity and inclusion, key strategies for promoting greater equity and representation, and the potential impact of inclusive practices on reducing systemic disparities and fostering shared prosperity.

198

Key Components and Characteristics:

Diverse Representation: Diverse representation entails ensuring that individuals from a variety of backgrounds, experiences, and perspectives are represented and included in decision-making processes, leadership positions, and workforce demographics. Diversity encompasses dimensions such as race, ethnicity, gender, sexual orientation, age, disability status, socioeconomic background, and cultural identity, reflecting the full spectrum of human diversity and enriching organizational and community dynamics.

Inclusive Culture and Environment:

Inclusive culture and environment create spaces where all individuals feel valued, respected, and empowered to contribute their unique talents, perspectives, and identities. Inclusive workplaces, schools, and communities foster a sense of belonging, psychological safety, and mutual respect among diverse stakeholders, enabling individuals to thrive, collaborate, and

199

succeed regardless of differences in background or identity.

Equitable Policies and Practices:
Equitable policies and practices promote fairness, transparency, and accountability in recruitment, hiring, promotion, compensation, and performance evaluation processes, ensuring that opportunities for advancement and recognition are accessible to all individuals on the basis of merit and qualifications rather than bias or discrimination. Equity-focused initiatives address systemic barriers and unconscious biases that perpetuate inequities and limit access to opportunities for underrepresented groups.

Diversity Training and Education:
Diversity training and education programs raise awareness, build cultural competence, and foster inclusive behaviors and attitudes among employees, leaders, and stakeholders. These programs provide opportunities for learning, dialogue, and reflection on issues related to diversity, equity, and inclusion, equipping participants with the knowledge, skills, and tools needed to navigate diverse

environments, address bias, and promote allyship and solidarity across differences.

Affinity Groups and Employee Resource Networks:
Affinity groups and employee resource networks provide forums for individuals with shared identities or interests to connect, support each other, and advocate for inclusion and representation within organizations and communities. These groups offer networking opportunities, mentorship programs, and professional development initiatives that promote belonging, career advancement, and leadership opportunities for members from underrepresented or marginalized backgrounds.

Supplier Diversity Programs:
Supplier diversity programs promote economic inclusion and equity by fostering relationships with diverse suppliers, vendors, and business partners from underrepresented communities. These programs prioritize procurement contracts, purchasing agreements, and supplier partnerships with minority-owned, women-owned, veteran-owned, LGBTQ+-owned, and small businesses, thereby expanding economic opportunities, promoting

entrepreneurship, and diversifying supply chains.

Inclusive Marketing and Branding:
Inclusive marketing and branding efforts reflect diverse perspectives, experiences, and identities in advertising, media representation, and brand messaging, resonating with diverse audiences and consumers and fostering a sense of belonging and connection. Inclusive marketing campaigns challenge stereotypes, celebrate diversity, and amplify underrepresented voices, contributing to greater social awareness, cultural representation, and market relevance.

Community Engagement and Partnerships:
Community engagement and partnerships involve collaborating with local stakeholders, community organizations, and advocacy groups to address systemic barriers, promote social justice, and advance equity and inclusion within neighborhoods, schools, workplaces, and civic institutions. These partnerships leverage collective expertise, resources, and networks to advocate for policy reforms, drive community-led

initiatives, and create inclusive environments where everyone can thrive.

Data Collection and Analysis:

Data collection and analysis efforts collect disaggregated demographic data, conduct diversity assessments, and track progress toward diversity and inclusion goals and benchmarks. Data-driven approaches enable organizations to identify disparities, measure the impact of diversity and inclusion initiatives, and make evidence-based decisions to address inequities, promote accountability, and drive continuous improvement in diversity and inclusion outcomes.

Leadership Commitment and Accountability:

Leadership commitment and accountability are essential for driving meaningful change and embedding diversity and inclusion principles into organizational culture, policies, and practices. Executive leadership, board members, and senior management teams play a critical role in setting the tone, defining priorities, allocating resources, and holding themselves and others accountable for advancing diversity, equity, and inclusion goals and outcomes.

Case Studies and Success Stories:

Salesforce: Salesforce is a technology company known for its commitment to diversity, equity, and inclusion (DEI) initiatives. The company has implemented a range of programs and policies to advance workforce diversity, including unconscious bias training, pay equity assessments, and employee resource groups focused on diversity, women, LGBTQ+, and other affinity groups. Salesforce's DEI efforts have resulted in increased representation of underrepresented groups in leadership positions and a more inclusive workplace culture.

The Walt Disney Company: The Walt Disney Company has prioritized diversity and inclusion across its entertainment brands, theme parks, and media platforms. Through initiatives such as the Disney Aspire education program, which provides tuition assistance and educational resources for employees, and diverse storytelling and content creation

efforts, Disney promotes representation, celebrates cultural diversity, and fosters inclusion within its workforce and content offerings.

Procter & Gamble: Procter & Gamble (P&G) is a consumer goods company committed to advancing diversity and inclusion in its global operations. P&G's Supplier Diversity Program promotes economic empowerment by sourcing from diverse suppliers and fostering supplier partnerships with minority-owned and women-owned businesses. The company also invests in diversity training, leadership development, and community engagement initiatives that promote equity and inclusion within its workforce and supply chain.

Conclusion:

Diversity and inclusion initiatives are essential drivers of social progress, economic equity, and shared prosperity in the United States. By promoting diversity, fostering inclusion, and advancing equity within workplaces, communities, and society, organizations and institutions can address systemic disparities, expand opportunities for underrepresented groups,

205

and create environments where everyone has the opportunity to thrive and succeed. Through collaborative efforts, transformative leadership, and a commitment to equity and justice, we can build a more inclusive and equitable economy that reflects the diversity of our communities and unlocks the full potential of all individuals.

Ethical Business Practices

Introduction: Ethical business practices serve as a cornerstone for addressing the wealth gap and fostering a more equitable economy in the United States. Ethical behavior in business encompasses principles of integrity, transparency, accountability, and social responsibility, guiding organizations to prioritize the well-being of stakeholders, communities, and the environment. This chapter explores the importance of ethical business practices, key principles for promoting integrity and social responsibility, and the potential impact of ethical conduct on reducing inequality and promoting shared prosperity.

Key Components and Characteristics:

Integrity and Honesty: Integrity and honesty are foundational principles of ethical business practices, guiding organizations to uphold truthfulness, fairness, and transparency in their dealings with stakeholders, customers, employees, and the public. Ethical leaders and employees demonstrate integrity by adhering to high ethical standards, fulfilling commitments, and acting in the best interests of all stakeholders, even when faced with difficult decisions or competing interests.

Compliance with Laws and Regulations: Compliance with laws and regulations ensures that businesses operate within legal boundaries and uphold legal standards of conduct in their business activities, transactions, and operations. Ethical organizations prioritize legal compliance, maintain accurate records, and adhere to industry-specific regulations, codes of conduct, and professional standards to prevent fraud, corruption, and unethical behavior.

207

Fair Treatment and Equal Opportunity: Fair treatment and equal opportunity promote equity, diversity, and inclusion within the workplace and beyond, ensuring that all individuals have access to employment, advancement, and opportunities for success regardless of factors such as race, ethnicity, gender, age, sexual orientation, disability, or socioeconomic status. Ethical businesses foster inclusive cultures, implement non-discrimination policies, and provide equitable access to resources, training, and career development opportunities for all employees.

Responsible Supply Chain Management: Responsible supply chain management practices promote ethical sourcing, environmental sustainability, and social responsibility throughout the supply chain, from sourcing raw materials to manufacturing, distribution, and retail. Ethical businesses prioritize suppliers and partners that adhere to fair labor practices, environmental regulations, and human rights standards, ensuring that products and services are produced ethically and sustainably without exploitation or harm to workers, communities, or the environment.

Environmental Sustainability and Stewardship: Environmental sustainability and stewardship practices promote responsible use of natural resources, minimize environmental impact, and mitigate climate change by adopting sustainable business practices, reducing carbon emissions, and conserving energy and water resources. Ethical businesses invest in renewable energy, green technologies, and eco-friendly practices that support environmental sustainability, reduce waste, and promote corporate citizenship and environmental stewardship.

Consumer Protection and Privacy: Consumer protection and privacy policies safeguard consumer rights, privacy, and data security by ensuring that products and services meet quality standards, adhere to safety regulations, and respect consumer privacy and confidentiality. Ethical businesses prioritize consumer well-being, provide accurate information, and protect consumer data from unauthorized access, disclosure, or misuse, fostering trust,

loyalty, and brand reputation among customers and stakeholders.

Corporate Social Responsibility (CSR) Initiatives:
Corporate social responsibility (CSR) initiatives enable businesses to contribute positively to society by investing in philanthropy, community development, and social impact projects that address pressing social, environmental, and economic challenges. Ethical businesses engage in CSR activities such as charitable giving, volunteerism, and community partnerships that promote education, healthcare, environmental conservation, and economic empowerment, creating shared value for communities, stakeholders, and society at large.

Transparency and Accountability:
Transparency and accountability are essential for building trust, credibility, and confidence among stakeholders and the public. Ethical businesses practice transparency by disclosing relevant information about their business practices, financial performance, governance structures, and social and environmental impacts to stakeholders,

investors, regulators, and the public. Accountability mechanisms, such as independent audits, stakeholder engagement, and governance oversight, ensure that businesses are held accountable for their actions and decisions and that corrective actions are taken when necessary to address deficiencies or wrongdoing.

Ethical Leadership and Corporate Governance: Ethical leadership and corporate governance set the tone for ethical behavior and organizational culture, guiding decision-making processes, values, and priorities within businesses and institutions. Ethical leaders demonstrate integrity, humility, and ethical courage by modeling ethical behavior, fostering open communication, and holding themselves and others accountable for upholding ethical standards and principles. Strong corporate governance structures, independent oversight mechanisms, and ethical leadership practices promote transparency, integrity, and accountability in business operations and decision-making processes.

211

Continuous Improvement and Learning: Continuous improvement and learning are integral to ethical business practices, encouraging organizations to reflect on past experiences, learn from mistakes, and embrace opportunities for growth and development. Ethical businesses foster a culture of learning, innovation, and continuous improvement by encouraging feedback, embracing diversity of thought, and adapting to changing social, economic, and environmental contexts to ensure that business practices remain ethical, relevant, and responsive to evolving stakeholder expectations and societal needs.

Case Studies and Success Stories:

Patagonia: Patagonia, an outdoor apparel company, is renowned for its commitment to environmental sustainability, social responsibility, and ethical business practices. The company invests in sustainable sourcing, fair labor practices, and environmental conservation initiatives, demonstrating its dedication to corporate citizenship and environmental stewardship.

212

Ben & Jerry's: Ben & Jerry's, an ice cream company, is known for its social activism, advocacy for social justice, and commitment to ethical business practices. The company champions fair trade sourcing, promotes social equity, and advocates for progressive causes through its business operations, marketing campaigns, and philanthropic initiatives, aligning its values with its business practices to create positive social impact.

Interface: Interface, a global flooring company, is a leader in sustainable business practices and corporate social responsibility. The company's Mission Zero commitment aims to eliminate negative environmental impacts from its operations, products, and supply chain by 2020, demonstrating its commitment to environmental sustainability, innovation, and corporate citizenship.

Conclusion:

Ethical business practices are essential for creating a more equitable economy that prioritizes integrity, transparency, and social responsibility. By embracing ethical principles and values, businesses can contribute to reducing the wealth gap, promoting social justice, and fostering shared prosperity for all stakeholders. Through ethical leadership, responsible business practices, and a commitment to continuous improvement, we can build a more ethical, inclusive, and sustainable economy that serves the needs of people, communities, and the planet.

CHAPTER 4

CHALLENGES AND OBSTACLES

Introduction: Addressing the wealth gap and striving towards a more equitable economy in the United States is a complex and multifaceted endeavor. While there are numerous solutions and strategies to pursue, there are also significant challenges and obstacles that must be acknowledged and overcome. This chapter explores some of the key challenges and obstacles that impede progress towards economic equity and shared prosperity, shedding light on the systemic barriers and structural inequalities that perpetuate the wealth gap.

Structural Inequality: Structural inequality refers to the systemic barriers and disparities embedded within social, economic, and political systems that perpetuate unequal access to opportunities, resources, and wealth accumulation. Historical injustices, discriminatory policies, and institutionalized racism have created

215

structural inequalities that disproportionately disadvantage marginalized communities, exacerbating the wealth gap and hindering upward mobility for generations.

Racial and Gender Disparities:

Racial and gender disparities are pervasive challenges that contribute to the wealth gap and perpetuate economic inequities in the United States. People of color and women face systemic barriers in education, employment, housing, and access to financial services, resulting in lower wages, limited wealth accumulation, and reduced economic mobility compared to their white, male counterparts.

Income Inequality:

Income inequality is a significant obstacle to achieving economic equity and shared prosperity. The concentration of wealth and income among the top echelons of society widens the wealth gap, exacerbates social inequality, and undermines economic stability and social cohesion. Rising income inequality limits opportunities for upward mobility, diminishes social mobility, and perpetuates intergenerational poverty and disadvantage.

Lack of Access to Quality Education: The lack of access to quality education perpetuates the wealth gap by limiting opportunities for educational attainment and economic advancement, particularly among low-income communities and communities of color. Disparities in educational resources, funding, and quality contribute to unequal educational outcomes, widening the achievement gap and hindering economic mobility for disadvantaged individuals and communities.

Systemic Racism and Discrimination: Systemic racism and discrimination continue to shape economic opportunities and outcomes for marginalized communities, perpetuating the wealth gap and hindering efforts to achieve economic equity. Racial biases in hiring, lending, housing, and criminal justice contribute to disparities in income, wealth, and access to opportunities, creating systemic barriers that prevent marginalized groups from realizing their full potential and participating fully in the economy.

Limited Access to Affordable Housing:

Limited access to affordable housing exacerbates economic inequality and contributes to the wealth gap by disproportionately burdening low-income households and limiting their ability to build wealth through homeownership and property ownership. Skyrocketing housing costs, gentrification, and discriminatory housing practices further marginalize vulnerable communities, exacerbating housing insecurity and perpetuating cycles of poverty and homelessness.

Barriers to Entrepreneurship and Small Business Ownership: Barriers to entrepreneurship and small business ownership hinder economic mobility and wealth creation, particularly among marginalized communities and historically disadvantaged groups. Limited access to capital, regulatory hurdles, lack of mentorship and support networks, and discriminatory lending practices inhibit entrepreneurship and small business growth, perpetuating economic disparities and limiting opportunities for wealth accumulation.

218

Inadequate Social Safety Net:

Inadequate social safety net programs fail to adequately address the needs of vulnerable populations and mitigate the impacts of economic hardship and inequality. Insufficient access to healthcare, childcare, paid leave, unemployment benefits, and other social supports exacerbates financial insecurity, perpetuates poverty, and widens the wealth gap, particularly during times of economic crisis and uncertainty.

Political Polarization and Policy Gridlock:

Political polarization and policy gridlock impede progress towards addressing the wealth gap and implementing effective solutions for economic equity. Partisan politics, special interests, and ideological divisions hinder consensus-building and policymaking, delaying critical reforms and exacerbating social and economic inequalities. Lack of political will and leadership commitment further hinder efforts to enact meaningful change and advance social justice.

Resistance to Change and Status Quo:

Resistance to change and

the status quo perpetuate the wealth gap by preserving existing power structures, economic systems, and institutional arrangements that benefit privileged groups at the expense of marginalized communities. Entrenched interests, vested stakeholders, and inertia hinder efforts to dismantle systemic barriers, challenge inequitable practices, and promote transformative change, perpetuating cycles of injustice and inequality.

Conclusion: Addressing the wealth gap and advancing economic equity in the United States requires confronting and overcoming numerous challenges and obstacles that perpetuate systemic inequalities and hinder progress towards shared prosperity. By acknowledging the structural barriers and systemic injustices that underlie the wealth gap, we can work towards implementing bold, equitable solutions that dismantle inequitable systems, promote social justice, and create a more inclusive and prosperous economy for all. Through collective action, advocacy, and commitment to justice, we can overcome these challenges and build a more equitable society where everyone has the opportunity to thrive and succeed.

220

Political Resistance

Introduction: Political resistance poses a significant obstacle to addressing the wealth gap and implementing solutions for creating a more equitable economy in the United States. Despite widespread recognition of the need for systemic change, political resistance from various quarters often undermines efforts to enact meaningful reforms and policies that promote economic justice and reduce inequality. This chapter examines the sources, motivations, and implications of political resistance to addressing the wealth gap, highlighting the challenges it presents and strategies for navigating and overcoming resistance to change.

Special Interest Influence:

Political resistance to addressing the wealth gap often stems from the influence of powerful special interest groups, including corporate lobbyists, industry associations, and wealthy donors, who seek to protect their vested interests and preserve the status quo. These special interests wield considerable influence over elected officials through campaign contributions, lobbying efforts, and policy

221

advocacy, shaping legislative agendas and priorities to serve their narrow interests at the expense of broader social and economic equity.

Partisan Politics: Partisan politics exacerbate political resistance to addressing the wealth gap by fostering ideological divisions and gridlock that hinder consensus-building and policymaking. Political polarization between Democrats and Republicans often leads to partisan gridlock, legislative stalemates, and ideological battles that prevent bipartisan cooperation on critical issues such as tax reform, income inequality, and social welfare policies, perpetuating systemic inequalities and hindering progress towards economic equity.

Ideological Opposition: Ideological opposition to government intervention and redistributive policies poses a formidable barrier to addressing the wealth gap, particularly among conservative policymakers and free-market proponents who advocate for limited government intervention and deregulation. Ideological opposition to progressive taxation, social

222

welfare programs, and wealth redistribution initiatives reflects deeply held beliefs in individualism, free markets, and limited government intervention, complicating efforts to enact policies that promote economic justice and reduce inequality.

Corporate Capture of Government:
Political resistance to addressing the wealth gap is exacerbated by the phenomenon of corporate capture, whereby powerful corporations and industry groups exert undue influence over government decision-making processes, regulatory agencies, and public policy outcomes. Regulatory capture, revolving door practices, and corporate lobbying efforts enable corporations to shape laws, regulations, and policies in their favor, undermining democratic governance and perpetuating economic disparities that benefit the corporate elite at the expense of ordinary citizens.

Campaign Finance Reform:
Campaign finance reform is essential for reducing political resistance to addressing the wealth gap and mitigating the influence of money in politics. Reforms such as

public financing of elections, contribution limits, transparency requirements, and restrictions on corporate and dark money contributions can help level the playing field, empower grassroots movements, and curb the influence of wealthy donors and special interests over political decision-making processes.

Voter Education and Mobilization: Voter education and mobilization efforts are critical for overcoming political resistance to addressing the wealth gap and building grassroots support for progressive policies that promote economic equity and social justice. Empowering voters with information about the impacts of wealth inequality, corporate influence, and policy choices on their lives and communities can inspire civic engagement, mobilize voter turnout, and hold elected officials accountable for advancing policies that serve the public interest.

Coalition Building and Advocacy: Coalition building and advocacy play a vital role in overcoming political resistance to addressing the wealth gap by mobilizing diverse

stakeholders, grassroots organizations, labor unions, advocacy groups, and community organizers to advocate for progressive policies and systemic reforms. Building broad-based coalitions, forging alliances across sectors and movements, and amplifying marginalized voices can strengthen advocacy efforts, amplify collective demands, and increase political pressure on policymakers to prioritize economic equity and social justice.

Legislative Strategy and Policy Innovation:

Legislative strategy and policy innovation are essential for overcoming political resistance and advancing progressive solutions to address the wealth gap. Proponents of economic justice must develop strategic approaches to navigate political obstacles, build consensus, and advance policy agendas that prioritize investments in education, healthcare, affordable housing, progressive taxation, and social welfare programs that promote equity and opportunity for all.

Public Awareness and Grassroots Organizing:

Public awareness and grassroots organizing efforts are crucial for overcoming political

225

resistance to addressing the wealth gap by raising awareness, mobilizing public support, and fostering a sense of urgency around the need for systemic change. Grassroots movements, social media campaigns, and community organizing initiatives can amplify marginalized voices, elevate public discourse, and pressure policymakers to prioritize economic equity and social justice in their policy agendas.

Electoral Accountability and Civic Engagement:

Electoral accountability and civic engagement are essential mechanisms for holding elected officials accountable and overcoming political resistance to addressing the wealth gap. Engaging in electoral politics, participating in local elections, and holding elected officials accountable for their actions and policy decisions can help ensure that political leaders represent the interests of their constituents, prioritize economic equity, and enact policies that promote shared prosperity and social well-being.

Conclusion:

Political resistance poses a formidable challenge to addressing the wealth gap and achieving economic equity in the United States. Overcoming political resistance requires concerted efforts to challenge entrenched power dynamics, mobilize grassroots movements, and build broad-based coalitions that advocate for progressive policies and systemic reforms. By empowering voters, amplifying marginalized voices, and holding elected officials accountable, we can overcome political resistance and advance policies that promote economic justice, reduce inequality, and create a more equitable economy for all.

Structural Inertia

Introduction: Structural inertia refers to the resistance to change inherent within existing social, economic, and political systems, which impedes efforts to address the wealth gap and create a more equitable economy in the United States. Despite the pressing need for systemic reforms and transformative policies, structural inertia perpetuates entrenched inequalities, preserves status quo power dynamics, and hinders progress towards economic justice and shared prosperity. This chapter examines the sources, manifestations, and consequences of structural inertia, as well

227

as strategies for overcoming resistance to change and fostering a more equitable economic system.

Entrenched Power Structures:

Structural inertia is reinforced by entrenched power structures that concentrate wealth, privilege, and influence among dominant elites and institutions, perpetuating systemic inequalities and limiting opportunities for marginalized communities to access resources, representation, and economic mobility. Concentrated power among economic elites, corporate interests, and political elites enables them to resist efforts to challenge the status quo, preserve their vested interests, and maintain their dominance over economic and political decision-making processes.

Institutional Resistance to Reform: Structural inertia is reinforced by institutional resistance to reform within existing social, economic, and political institutions, which are often resistant to change and reluctant to challenge established norms, practices, and power dynamics. Bureaucratic inertia, regulatory capture, and institutional barriers hinder efforts to enact meaningful reforms and

228

policies that promote economic equity, perpetuating systemic inequalities and limiting the effectiveness of government interventions and social programs.

Path Dependence and Historical Legacies:

Structural inertia is perpetuated by path dependence and historical legacies that shape existing institutions, policies, and social structures, limiting the range of feasible options for addressing the wealth gap and creating a more equitable economy. Historical injustices, discriminatory policies, and systemic biases embedded within social and economic systems create inertia that perpetuates existing patterns of inequality and impedes efforts to dismantle systemic barriers and create more inclusive and equitable institutions.

Cultural Norms and Beliefs:

Structural inertia is reinforced by cultural norms, beliefs, and narratives that rationalize and justify existing inequalities, perpetuating social hierarchies, stereotypes, and stigmatization of marginalized groups. Cultural narratives that valorize individualism, meritocracy, and rugged self-reliance legitimize socioeconomic disparities, undermine collective solidarity, and hinder efforts to build inclusive, equitable societies that prioritize the well-being and dignity of all individuals.

Inequality-Reinforcing Feedback Loops: Structural inertia is perpetuated by inequality-reinforcing feedback loops that amplify existing disparities and entrench privilege and disadvantage over time. Inequality begets further inequality as wealth and power concentrate among the privileged few, while marginalized communities face systemic barriers to accessing opportunities, resources, and social mobility, perpetuating cycles of poverty, exclusion, and intergenerational disadvantage.

Lack of Political Will and Leadership: Structural inertia is reinforced by a lack of political will and leadership commitment to addressing the wealth gap and enacting transformative reforms that challenge existing power structures and privilege. Political leaders may prioritize short-term political expediency over long-term systemic change, succumb to pressure from special

230

interests and powerful stakeholders, or lack the vision, courage, and commitment to champion bold, equitable policies and initiatives that promote economic justice and shared prosperity.

Resistance to Redistribution and Economic Reforms: Structural inertia is reinforced by resistance to redistribution and economic reforms that challenge vested interests, wealth accumulation, and economic privilege. Efforts to enact progressive taxation, wealth redistribution, and social welfare programs are often met with opposition from economic elites, corporate interests, and conservative policymakers who advocate for laissez-faire policies, trickle-down economics, and deregulation, perpetuating structural inequalities and hindering efforts to create a more equitable economy.

Social and Economic Fragmentation: Structural inertia is exacerbated by social and economic fragmentation that weakens social cohesion, solidarity, and collective action, making it difficult to mobilize broad-based support for progressive policies and systemic reforms. Fragmentation along

231

racial, ethnic, class, and ideological lines undermines efforts to build coalitions, forge alliances, and mobilize grassroots movements that challenge existing power structures and advocate for inclusive, equitable policies that benefit all members of society.

Corporate Influence and Capture of Government:
Structural inertia is reinforced by corporate influence and capture of government institutions, which enable powerful corporations and industry groups to shape public policies, regulatory frameworks, and legislative agendas in ways that serve their narrow interests and perpetuate systemic inequalities. Regulatory capture, corporate lobbying, and revolving door practices undermine democratic governance, erode public trust, and hinder efforts to enact policies that promote economic justice, environmental sustainability, and social welfare.

Strategic Interventions and Systemic Reforms:
Overcoming structural inertia requires strategic interventions and systemic reforms that challenge existing power structures, dismantle institutional barriers, and foster

232

inclusive, equitable institutions and policies. Proponents of economic justice must advocate for transformative change, build broad-based coalitions, and mobilize grassroots movements to challenge entrenched interests, shift cultural norms, and create political momentum for progressive policies and initiatives that prioritize the well-being and dignity of all individuals.

Conclusion: Structural inertia poses formidable challenges to addressing the wealth gap and creating a more equitable economy in the United States. Overcoming structural inertia requires collective action, visionary leadership, and sustained advocacy for systemic reforms that challenge existing power structures, dismantle institutional barriers, and foster inclusive, equitable institutions and policies. By confronting structural inertia head-on and mobilizing collective efforts to build a more just, inclusive, and equitable society, we can create a future where everyone has the opportunity to thrive and participate fully in the economic, social, and political life of our nation.

Cultural Attitudes and Beliefs

Introduction: Cultural attitudes and beliefs play a significant role in shaping perceptions of wealth, poverty, and economic inequality in the United States. Cultural narratives, values, and norms influence individual behaviors, societal expectations, and public policies related to economic opportunity, social mobility, and wealth distribution. Understanding cultural attitudes and beliefs is essential for addressing the wealth gap and creating a more equitable economy that reflects shared values of fairness, justice, and opportunity for all. This chapter explores the cultural attitudes and beliefs that contribute to the wealth gap, examines their impact on social and economic outcomes, and identifies strategies for shifting cultural narratives towards greater equity and inclusion.

Individualism and Meritocracy:

The cultural belief in individualism and meritocracy reinforces the notion that personal success and wealth accumulation are primarily determined by individual

234

effort, talent, and hard work. The American Dream, rooted in the belief that anyone can achieve prosperity through diligence and determination, fosters a culture of self-reliance and upward mobility. While individualism and meritocracy celebrate achievement and ambition, they also obscure systemic barriers and structural inequalities that limit opportunities for marginalized communities and perpetuate the wealth gap.

Wealth as a Marker of Success:
Cultural attitudes often equate wealth with success, status, and achievement, reinforcing the idea that material wealth is a measure of personal worth and social standing. The glorification of wealth in popular culture, media, and advertising perpetuates consumerism, materialism, and status-seeking behaviors that prioritize accumulation and conspicuous consumption over collective well-being and social responsibility. The cultural valorization of wealth perpetuates inequalities and exacerbates the wealth gap by perpetuating social hierarchies and reinforcing privilege and disadvantage.

Stigmatization of Poverty and Economic Struggle: Cultural attitudes towards poverty and economic struggle often stigmatize individuals and communities experiencing financial hardship, blaming them for their circumstances and reinforcing negative stereotypes of laziness, irresponsibility, and moral failure. The cultural stigma attached to poverty and economic insecurity perpetuates social exclusion, marginalization, and discrimination against low-income individuals and communities, hindering efforts to address the root causes of poverty and promote economic justice.

Consumerism and Materialism: Cultural attitudes towards consumerism and materialism prioritize material wealth, possessions, and consumption as symbols of success, happiness, and fulfillment. The culture of consumerism encourages excessive spending, debt accumulation, and overconsumption of goods and services, fueling unsustainable patterns of consumption and exacerbating inequalities in wealth distribution. The emphasis on material wealth as a source of happiness

and fulfillment perpetuates socioeconomic disparities and undermines efforts to prioritize collective well-being and social equity.

Inequality and Social Division:
Cultural attitudes towards inequality and social division shape perceptions of wealth, privilege, and social status, influencing attitudes towards redistribution, taxation, and social welfare policies. Cultural narratives that emphasize rugged individualism and personal responsibility often downplay the role of structural factors and systemic injustices in perpetuating the wealth gap, fostering resentment, mistrust, and polarization among different social groups. The normalization of inequality and social division undermines solidarity, cohesion, and collective action needed to address systemic inequalities and promote economic justice.

Cultural Narratives of Success and Failure:
Cultural narratives of success and failure shape attitudes towards achievement, resilience, and adversity, influencing perceptions of individuals' abilities to overcome barriers

237

and achieve economic success. Cultural narratives that celebrate triumph over adversity often overlook systemic barriers and structural inequalities that limit opportunities for marginalized communities and perpetuate intergenerational poverty and disadvantage. The glorification of individual success narratives can obscure the structural factors and systemic injustices that perpetuate the wealth gap, reinforcing cultural myths of meritocracy and self-reliance.

Cultural Values of Work and Productivity: Cultural values of work and productivity emphasize the importance of labor, diligence, and productivity as pathways to economic security and social mobility. The Protestant work ethic, rooted in the belief that hard work and industriousness are moral virtues, shapes cultural attitudes towards labor and wealth accumulation, reinforcing the idea that success is earned through effort and perseverance. While cultural values of work and productivity celebrate industriousness and ambition, they also obscure systemic barriers and structural inequalities that limit opportunities for

economic advancement and perpetuate the wealth gap.

Cultural Diversity and Inclusion:

Cultural attitudes towards diversity and inclusion influence perceptions of social identity, belonging, and representation within society. Cultures that celebrate diversity, equity, and inclusion prioritize representation, respect, and recognition of diverse perspectives, experiences, and identities. Embracing cultural diversity and promoting inclusive cultural narratives can challenge stereotypes, combat discrimination, and foster social cohesion, empowering individuals and communities to participate fully in economic, political, and social life.

Conclusion:

Cultural attitudes and beliefs profoundly shape perceptions of wealth, poverty, and economic inequality in the United States, influencing individual behaviors, societal values, and public policies related to economic opportunity and social mobility. By recognizing the impact of cultural attitudes and beliefs on the wealth gap and promoting narratives of equity, justice, and opportunity for all, we can challenge existing norms, shift cultural

239

narratives, and build a more inclusive and equitable economy that reflects shared values of fairness, dignity, and prosperity for all.

Implementation Challenges

Introduction: Addressing the wealth gap and creating a more equitable economy in the United States requires navigating numerous implementation challenges that arise from complex social, economic, and political dynamics. While there is widespread recognition of the need for systemic reforms and transformative policies, translating intentions into tangible actions and outcomes poses formidable obstacles. This chapter examines the implementation challenges associated with addressing the wealth gap, explores their root causes, and identifies strategies for overcoming barriers to effective implementation.

Political Resistance and Partisan Gridlock: Political resistance and partisan gridlock present significant challenges to implementing policies and reforms aimed at addressing the wealth gap. Ideological divisions,

240

partisan politics, and special interest influence can hinder bipartisan cooperation, legislative consensus, and policy implementation, delaying critical reforms and perpetuating systemic inequalities.

Lack of Funding and Resources:

Insufficient funding and resources pose barriers to implementing programs and initiatives designed to reduce the wealth gap and promote economic equity. Budgetary constraints, competing priorities, and fiscal austerity measures may limit investments in education, social welfare programs, affordable housing, and infrastructure development, hindering efforts to address systemic barriers and create pathways to opportunity for marginalized communities.

Complexity and Interconnectedness of Issues:

The complexity and interconnectedness of issues related to the wealth gap pose implementation challenges, as systemic inequalities cut across multiple domains, including education, employment, housing, healthcare, and criminal justice. Addressing the wealth gap requires holistic, multi-sectoral approaches that recognize the interplay of social, economic, and political factors shaping outcomes and opportunities for individuals and communities.

Institutional Inertia and Bureaucratic Barriers: Institutional inertia and bureaucratic barriers within existing systems can impede efforts to implement reforms and initiatives aimed at reducing the wealth gap. Resistance to change, entrenched power structures, and administrative hurdles may slow the pace of implementation, stifle innovation, and limit the effectiveness of interventions designed to promote economic justice and shared prosperity.

Resistance from Entrenched Interests: Resistance from entrenched interests, including corporate lobbyists, industry groups, and wealthy donors, can undermine efforts to implement policies and reforms that challenge existing power structures and privilege. Special interest influence, regulatory capture, and corporate lobbying efforts may thwart attempts to enact progressive taxation, financial regulation, and social welfare

programs that promote economic equity and social justice.

Data Limitations and Measurement Challenges:

Data limitations and measurement challenges complicate efforts to assess progress, evaluate outcomes, and track the impact of interventions aimed at reducing the wealth gap. Incomplete data, methodological limitations, and disparities in data collection and reporting may obscure patterns of inequality, hinder evidence-based decision-making, and undermine accountability for achieving equitable outcomes.

Public Perception and Cultural Attitudes:

Public perception and cultural attitudes towards wealth, poverty, and economic inequality can shape receptivity to policy interventions and influence the success of implementation efforts. Misconceptions, stigma, and ideological biases may fuel resistance to progressive policies, perpetuate stereotypes, and undermine support for initiatives aimed at reducing the wealth gap and promoting economic justice.

Geographic Disparities and Regional Inequities: Geographic disparities and regional inequities exacerbate implementation challenges by exacerbating inequalities in access to resources, opportunities, and social services. Rural-urban divides, regional disparities in economic development, and uneven distribution of infrastructure and investment may exacerbate the wealth gap and hinder efforts to promote inclusive growth and shared prosperity across diverse communities.

Capacity Constraints and Human Resources: Capacity constraints and human resource limitations can constrain the ability of governments, non-profit organizations, and community groups to implement effective strategies for addressing the wealth gap. Shortages of skilled personnel, insufficient training, and turnover in key positions may undermine organizational capacity, impede coordination, and hamper efforts to scale up interventions and sustain long-term impact.

Community Engagement and Stakeholder Collaboration:

Community engagement and stakeholder collaboration are essential for effective implementation of initiatives aimed at addressing the wealth gap. Building trust, fostering partnerships, and engaging affected communities in decision-making processes can enhance program relevance, improve outcomes, and promote ownership and sustainability of interventions designed to promote economic equity and social inclusion.

Conclusion:

Addressing the wealth gap and creating a more equitable economy in the United States requires overcoming numerous implementation challenges that arise from complex social, economic, and political dynamics. By recognizing the root causes of implementation challenges and adopting strategies for overcoming barriers to effective implementation, policymakers, practitioners, and advocates can advance policies and initiatives that promote economic justice, reduce inequality, and create pathways to opportunity for all individuals and communities. Through collective action, innovation, and

245

commitment to systemic change, we can build a more inclusive and equitable society where everyone has the opportunity to thrive and succeed.

CHAPTER 5

THE ROLE OF GOVERNMENT, CIVIL SOCIETY, AND BUSINESSES

Introduction: Addressing the wealth gap and creating a more equitable economy in the United States requires coordinated action and collaboration among government institutions, civil society organizations, and businesses. Each sector plays a distinct yet interconnected role in shaping policies, mobilizing resources, and fostering collective efforts to promote economic justice, reduce inequality, and create pathways to opportunity for all. This chapter explores the roles of government, civil society, and businesses in addressing the wealth gap, examines their respective strengths and challenges, and identifies strategies for effective collaboration and partnership.

Government:

247

Policy Formulation and Regulation:
Government institutions are responsible for formulating and implementing policies that promote economic equity, social welfare, and inclusive growth. Through legislation, regulation, and public investment, governments can enact progressive taxation, social welfare programs, and economic reforms that address the root causes of the wealth gap and promote shared prosperity.

Resource Allocation and Redistribution:
Governments play a critical role in allocating resources, redistributing wealth, and addressing socioeconomic disparities through progressive taxation, public spending, and targeted investments in education, healthcare, housing, and social services. By prioritizing equitable resource allocation and social investment, governments can mitigate the impacts of economic inequality and expand opportunities for marginalized communities.

Promotion of Economic Inclusion:

Government policies and programs can promote economic inclusion by expanding access to education, healthcare, affordable housing, and financial services, particularly for low-income individuals and communities. By investing in human capital development, workforce training, and community development initiatives, governments can empower individuals and communities to participate fully in the economy and realize their full potential.

Civil Society:

Advocacy and Public Awareness:

Civil society organizations play a vital role in advocating for policies and reforms that promote economic justice, reduce inequality, and address the root causes of the wealth gap. Through grassroots organizing, public education campaigns, and policy advocacy, civil society can mobilize public support, raise awareness about systemic injustices, and hold government and business accountable for advancing equitable policies and practices.

249

Service Delivery and Community Empowerment:

Civil society organizations deliver essential services, programs, and resources to marginalized communities, including education, healthcare, housing, and social assistance. By providing direct support, building community capacity, and fostering grassroots empowerment, civil society can address immediate needs, build social capital, and promote collective action towards economic equity and social justice.

Bridge Building and Collaboration:

Civil society serves as a bridge between government, businesses, and communities, facilitating dialogue, collaboration, and partnership to address complex social and economic challenges. By fostering partnerships, facilitating multi-stakeholder initiatives, and promoting cross-sectoral collaboration, civil society can leverage diverse perspectives, resources, and expertise to drive collective impact and sustainable change.

Businesses:

Corporate Social Responsibility:

Businesses have a responsibility to address the wealth gap and promote economic equity through corporate social responsibility (CSR) initiatives, ethical business practices, and sustainable development strategies. By adopting inclusive hiring practices, fair wages, and employee benefits, businesses can enhance economic opportunity, reduce income inequality, and foster inclusive growth within their organizations and communities.

Innovation and Economic Development:

Businesses play a critical role in driving innovation, economic development, and job creation, particularly in underserved and marginalized communities. By investing in research and development, entrepreneurship, and small business development, businesses can stimulate economic growth, create employment opportunities, and foster economic resilience in communities disproportionately affected by the wealth gap.

251

Stakeholder Engagement and Impact Investing: Businesses can engage stakeholders, including investors, employees, customers, and communities, in decision-making processes and corporate governance structures to promote transparency, accountability, and social impact. Through impact investing, corporate philanthropy, and community partnerships, businesses can align financial interests with social and environmental goals, drive positive change, and contribute to building a more equitable and sustainable economy.

Conclusion: The roles of government, civil society, and businesses are interdependent and complementary in addressing the wealth gap and creating a more equitable economy in the United States. By leveraging their respective strengths, resources, and expertise, these sectors can collaborate effectively to advance policies, programs, and initiatives that promote economic justice, reduce inequality, and create pathways to opportunity for all individuals and communities. Through collective action, collaboration, and commitment to shared

252

values of fairness, inclusion, and prosperity, government, civil society, and businesses can build a more just, equitable, and sustainable future for generations to come.

Collaborative Approaches and Partnerships

Introduction: Addressing the wealth gap and creating a more equitable economy in the United States requires collaborative approaches and partnerships that leverage the expertise, resources, and networks of diverse stakeholders across government, civil society, businesses, and communities. Collaboration fosters innovation, builds consensus, and amplifies collective impact by bringing together stakeholders with complementary strengths, perspectives, and resources to tackle complex social and economic challenges. This chapter explores collaborative approaches and partnerships for addressing the wealth gap, examines successful models of collaboration, and identifies strategies for fostering effective partnerships.

Multi-Sectoral Partnerships:

Multi-sectoral partnerships involve collaboration among government agencies, civil society organizations, businesses, and community stakeholders to address the wealth gap and promote economic equity. By harnessing the collective expertise, resources, and networks of diverse stakeholders, multi-sectoral partnerships can drive systemic change, scale innovative solutions, and address root causes of inequality.

Cross-Sectoral Collaboration:

Cross-sectoral collaboration involves collaboration across different industries, sectors, and disciplines to address complex social and economic challenges associated with the wealth gap. By fostering dialogue, sharing best practices, and leveraging complementary strengths, cross-sectoral collaboration can catalyze innovation, promote knowledge exchange, and generate sustainable solutions that benefit communities and economies.

Public-Private Partnerships (PPPs):

Public-private partnerships (PPPs) involve collaboration between government entities

254

and private sector organizations to deliver
public services, finance infrastructure
projects, and address social and economic
challenges. PPPs can leverage private
sector expertise, innovation, and financing
to complement government resources and
enhance the efficiency, effectiveness, and
sustainability of initiatives aimed at
reducing the wealth gap.

Community-Led Initiatives:

Community-led initiatives empower local
communities to drive change, build
resilience, and address the wealth gap
from the grassroots level. By engaging
community members as active participants
and decision-makers, community-led
initiatives can foster ownership, promote
inclusion, and tailor solutions to meet the
unique needs and priorities of diverse
communities.

Collaborative Governance Structures:

Collaborative governance structures
involve the establishment of inclusive,
participatory decision-making processes
that engage stakeholders from
government, civil society, businesses, and
communities in shaping policies, programs,

and initiatives to address the wealth gap. By fostering transparency, accountability, and trust, collaborative governance structures can build consensus, mitigate conflicts, and ensure that diverse voices are heard and represented in decision-making processes.

Data Sharing and Knowledge Exchange:

Data sharing and knowledge exchange facilitate collaboration by promoting information sharing, research collaboration, and evidence-based decision-making among stakeholders. By sharing data, research findings, and best practices, stakeholders can enhance understanding of the root causes of the wealth gap, identify effective interventions, and inform policy development and implementation efforts.

Capacity Building and Technical Assistance:

Capacity building and technical assistance initiatives support stakeholders in developing the skills, knowledge, and resources needed to effectively address the wealth gap and promote economic equity. By providing training, mentoring, and technical support, capacity building initiatives can strengthen organizational capacity, foster leadership development, and empower stakeholders to implement effective strategies and initiatives.

Participatory Monitoring and Evaluation:

Participatory monitoring and evaluation involve engaging stakeholders in the monitoring, evaluation, and learning processes to assess the effectiveness, impact, and outcomes of collaborative initiatives aimed at addressing the wealth gap. By soliciting feedback, conducting evaluations, and adapting strategies based on lessons learned, stakeholders can improve accountability, transparency, and performance, and enhance the sustainability of collaborative efforts over time.

Conclusion: Collaborative approaches and partnerships are essential for addressing the wealth gap and creating a more equitable economy in the United States. By fostering multi-sectoral collaboration, cross-sectoral partnerships, and community-led initiatives, stakeholders can leverage their collective strengths,

257

resources, and expertise to drive systemic change, scale innovative solutions, and advance economic equity and social justice for all individuals and communities. Through inclusive, participatory approaches to governance, data sharing, capacity building, and evaluation, stakeholders can build trust, foster collaboration, and mobilize collective action to create a more just, inclusive, and sustainable future for generations to come.

Advocacy and Activism

Introduction: Advocacy and activism are essential drivers of change in addressing the wealth gap and fostering a more equitable economy in the United States. Advocates and activists play a crucial role in raising awareness, mobilizing communities, and advocating for policies and reforms that promote economic justice, reduce inequality, and create opportunities for all individuals and communities. This chapter explores the power of advocacy and activism in addressing the wealth gap, examines successful advocacy campaigns and movements, and identifies strategies for effective advocacy and grassroots organizing.

Raising Awareness and Building Coalitions:

Advocacy and activism raise awareness about the root causes and consequences of the wealth gap, mobilize support, and build coalitions across diverse communities, constituencies, and movements. By amplifying voices, sharing stories, and fostering solidarity, advocates and activists can galvanize public opinion, mobilize grassroots support, and drive momentum for policy change and social transformation.

Policy Advocacy and Reform:

Advocacy and activism drive policy change and reform by advocating for legislative and regulatory measures that address systemic barriers, promote economic equity, and expand opportunities for marginalized communities. Through grassroots organizing, lobbying, and direct action, advocates and activists can hold policymakers accountable, shape public discourse, and influence decision-making processes to advance policies and reforms that prioritize the needs and interests of all individuals and communities.

Community Organizing and Empowerment:

Advocacy and activism empower communities to advocate for their rights, demand accountability, and mobilize collective action to address the wealth gap and systemic injustices. By building grassroots power, organizing campaigns, and fostering leadership development, advocates and activists can empower individuals and communities to become agents of change, challenge entrenched power structures, and advocate for policies and practices that promote economic justice and social equity.

Intersectional Approaches to Advocacy:

Advocacy and activism adopt intersectional approaches that recognize the interconnectedness of race, gender, class, and other forms of identity in shaping experiences of economic inequality and marginalization. By centering the voices and experiences of marginalized communities, advocating for inclusive policies, and addressing

260

intersecting forms of oppression and discrimination, advocates and activists can advance more holistic and equitable solutions to the wealth gap and promote social justice for all.

Media and Communications Advocacy:

Advocacy and activism leverage media and communications strategies to amplify marginalized voices, shape public narratives, and challenge dominant discourses about wealth, poverty, and economic inequality. Through storytelling, media campaigns, and digital advocacy tools, advocates and activists can raise awareness, mobilize public support, and influence public opinion to build momentum for policy change and social transformation.

Legal Advocacy and Litigation:

Advocacy and activism engage in legal advocacy and litigation to challenge discriminatory practices, advocate for legal protections, and hold accountable institutions and individuals responsible for perpetuating the wealth gap and systemic injustices. By partnering with legal experts, filing lawsuits, and advocating for policy reforms, advocates and activists can secure legal victories, establish precedents, and advance the rights and interests of marginalized communities in the pursuit of economic justice and equality under the law.

Educational Advocacy and Outreach:

Advocacy and activism engage in educational advocacy and outreach to promote economic literacy, civic engagement, and grassroots organizing skills among individuals and communities affected by the wealth gap. By providing education, training, and resources, advocates and activists can empower individuals to understand their rights, advocate for change, and participate in democratic processes to shape policies and institutions that affect their lives.

Conclusion: Advocacy and activism are powerful catalysts for change in addressing the wealth gap and creating a more equitable economy in the United States. By raising awareness, driving policy change, empowering communities, and challenging systemic injustices, advocates and activists can mobilize collective action, build grassroots power, and advance economic justice and social equity for all individuals and communities. Through strategic advocacy campaigns, intersectional approaches, and collaborative partnerships, advocates and activists can harness the power of collective action to create a more just, inclusive, and sustainable future for generations to come.

Corporate Social Responsibility

Introduction: Corporate Social Responsibility (CSR) represents a critical dimension of addressing the wealth gap and fostering a more equitable economy in the United States. As corporations wield significant economic power and influence, their engagement in CSR initiatives can drive positive social impact, promote economic inclusion, and address systemic inequalities. This chapter explores the concept of CSR, examines its role in addressing the wealth gap, and identifies strategies for businesses to integrate CSR principles into their operations and practices.

263

Defining Corporate Social Responsibility:

Corporate Social Responsibility (CSR) refers to the ethical and sustainable business practices that corporations adopt to balance economic objectives with social and environmental considerations. CSR encompasses a range of activities, including philanthropy, community engagement, environmental stewardship, ethical labor practices, and diversity and inclusion initiatives, aimed at creating shared value for stakeholders and society at large.

Promoting Economic Inclusion and Opportunity:

CSR initiatives can promote economic inclusion and opportunity by expanding access to employment, education, training, and financial services for marginalized communities. By adopting inclusive hiring practices, offering job training and skills development programs, and supporting minority-owned businesses and entrepreneurs, corporations can create pathways to economic mobility and empowerment for underserved populations.

Investing in Community Development:

CSR involves investing in community development projects and initiatives that address the root causes of poverty, inequality, and social exclusion. By funding affordable housing initiatives, supporting educational programs, and investing in infrastructure development in low-income communities, corporations can contribute to the economic revitalization and social well-being of underserved neighborhoods and regions.

Environmental Sustainability and Stewardship:

CSR encompasses environmental sustainability and stewardship practices that minimize the environmental impact of corporate operations and promote sustainable resource management. By reducing carbon emissions, conserving natural resources, and adopting renewable energy sources, corporations can mitigate climate change, protect ecosystems, and promote environmental justice for communities disproportionately affected by environmental degradation.

265

Ethical Supply Chain Management:

CSR involves ensuring ethical supply chain management practices that prioritize fair labor standards, human rights, and responsible sourcing of materials and products. By implementing transparent supply chain practices, conducting ethical audits, and partnering with suppliers committed to fair labor practices, corporations can combat forced labor, child labor, and other forms of exploitation in global supply chains.

Stakeholder Engagement and Transparency:

CSR entails engaging stakeholders, including employees, customers, investors, and communities, in decision-making processes and fostering transparency and accountability in corporate governance. By soliciting feedback, addressing stakeholder concerns, and disclosing corporate performance on social and environmental metrics, corporations can build trust, enhance reputation, and demonstrate commitment to responsible business practices.

Measuring Impact and Accountability:

CSR requires corporations to measure the impact of their social and environmental initiatives and hold themselves accountable for achieving meaningful outcomes. By adopting rigorous impact assessment frameworks, tracking key performance indicators, and reporting on progress towards CSR goals, corporations can assess the effectiveness of their initiatives, identify areas for improvement, and demonstrate accountability to stakeholders and society.

Collaborative Partnerships and Collective Action:

CSR initiatives often involve collaborative partnerships and collective action among corporations, governments, civil society organizations, and communities to address complex social and economic challenges. By leveraging collective resources, expertise, and networks, corporations can amplify their impact, scale innovative solutions, and drive systemic change that advances economic equity, social justice, and sustainable development.

267

Conclusion: Corporate Social Responsibility (CSR) represents a powerful mechanism for corporations to contribute to addressing the wealth gap and creating a more equitable economy in the United States. By integrating CSR principles into their operations and practices, corporations can promote economic inclusion, invest in community development, advance environmental sustainability, and foster stakeholder engagement and accountability. Through strategic CSR initiatives, corporations can harness their influence and resources to drive positive social impact, address systemic inequalities, and build a more just and sustainable future for all individuals and communities.

Policy Advocacy and Legislative Action

Introduction: Policy advocacy and legislative action are indispensable tools for addressing the wealth gap and fostering a more equitable economy in the United States. Through strategic advocacy efforts and engagement with policymakers, advocates can shape public policies, influence legislative agendas, and

268

advocate for reforms that prioritize economic justice, reduce inequality, and create opportunities for all individuals and communities. This chapter explores the role of policy advocacy and legislative action in addressing the wealth gap, examines successful advocacy campaigns, and identifies strategies for effective advocacy and policy reform.

Understanding the Policy Landscape:

Effective policy advocacy begins with a comprehensive understanding of the policy landscape, including existing laws, regulations, and institutional frameworks that shape economic outcomes and opportunities. Advocates must analyze policy trends, identify systemic barriers, and assess the impact of existing policies on the wealth gap and economic inequality to inform advocacy strategies and priorities.

Building Coalitions and Alliances:

Building coalitions and alliances is essential for amplifying advocacy efforts, mobilizing diverse constituencies, and building political momentum for policy change. Advocates must forge strategic partnerships with grassroots organizations, labor unions, community groups, and other stakeholders to build collective power, leverage resources, and advance shared policy goals that address the root causes of the wealth gap.

Crafting Policy Proposals and Solutions:

Crafting policy proposals and solutions requires rigorous research, analysis, and collaboration to develop evidence-based policy recommendations that address systemic inequalities and promote economic equity. Advocates must engage stakeholders, solicit input from affected communities, and work with experts to design policies and reforms that prioritize the needs and interests of marginalized populations and promote inclusive growth.

Engaging with Policymakers and Stakeholders:

Engaging with policymakers and stakeholders is essential for advancing policy advocacy goals and building support for legislative action. Advocates must cultivate relationships with elected officials, government agencies, and decision-makers, and educate them about

270

the importance of addressing the wealth gap and promoting economic justice through policy reforms. By organizing meetings, testifying at hearings, and participating in policy forums, advocates can influence decision-making processes and shape policy outcomes.

Mobilizing Grassroots Support and Civic Engagement:

Mobilizing grassroots support and civic engagement is critical for building political pressure, raising public awareness, and mobilizing communities to advocate for policy change. Advocates must engage in grassroots organizing, mobilize constituents, and empower individuals to become active participants in the democratic process by contacting elected officials, participating in advocacy campaigns, and exercising their right to vote.

Navigating the Legislative Process:

Navigating the legislative process requires a nuanced understanding of legislative procedures, timelines, and decision-making dynamics at the local, state, and federal levels. Advocates must track legislation,

271

monitor committee hearings, and identify strategic opportunities for advocacy and intervention to advance policy priorities and overcome obstacles to legislative action.

Leveraging Media and Communications Strategies:

Leveraging media and communications strategies is essential for amplifying advocacy messages, shaping public opinion, and garnering media coverage for policy advocacy efforts. Advocates must develop compelling narratives, engage with journalists and media outlets, and utilize digital platforms and social media to raise awareness about the wealth gap, highlight policy solutions, and mobilize public support for legislative action.

Monitoring Implementation and Accountability:

Monitoring implementation and accountability is crucial for ensuring that policy reforms are effectively implemented and that government agencies and institutions uphold their commitments to addressing the wealth gap and promoting economic equity. Advocates must track policy implementation, collect data on

272

outcomes and impact, and hold policymakers and stakeholders accountable for delivering on their promises and commitments to advancing economic justice and social inclusion.

Conclusion: Policy advocacy and legislative action are powerful tools for driving systemic change, promoting economic justice, and addressing the wealth gap in the United States. By engaging in strategic advocacy efforts, building coalitions, crafting evidence-based policy solutions, and mobilizing grassroots support, advocates can influence legislative agendas, shape public policies, and create an enabling environment for equitable economic growth and shared prosperity. Through sustained advocacy and collective action, advocates can advance policies and reforms that prioritize the needs of marginalized communities, dismantle systemic barriers, and create pathways to opportunity for all individuals and families striving for a better future.

Case Studies and Success Stories

Introduction: Case studies and success stories provide valuable insights into

273

effective strategies, innovative approaches, and impactful initiatives aimed at addressing the wealth gap and fostering a more equitable economy in the United States. By examining real-world examples of successful interventions, policymakers, advocates, and practitioners can learn from best practices, identify replicable models, and leverage lessons learned to inform policy development, program design, and advocacy efforts. This chapter presents a selection of case studies and success stories that illustrate diverse approaches to addressing the wealth gap and promoting economic equity across different sectors and communities.

Community Wealth Building Initiatives:

Case Study: The Evergreen Cooperatives (Cleveland, Ohio)

The Evergreen Cooperatives is a network of employee-owned businesses in Cleveland, Ohio, established to create jobs, build wealth, and revitalize low-income neighborhoods. Through partnerships with anchor institutions, including hospitals and universities, the cooperatives provide employment opportunities, training, and

274

ownership stakes to residents, fostering economic inclusion and community wealth building.

Financial Inclusion and Asset Building Programs:

Case Study: The Earned Income Tax Credit (EITC)

The Earned Income Tax Credit (EITC) is a federal program that provides refundable tax credits to low- and moderate-income working individuals and families, helping them build assets, reduce poverty, and improve financial stability. Research shows that the EITC has lifted millions of families out of poverty and increased workforce participation, demonstrating the effectiveness of targeted income supports in addressing the wealth gap.

Affordable Housing and Homeownership Initiatives:

Case Study: Community Land Trusts (CLTs) Community Land Trusts (CLTs) are nonprofit organizations that acquire and steward land to provide affordable housing and preserve community assets. CLTs offer long-term affordable homeownership opportunities, protect against displacement, and promote wealth accumulation for low-income residents, particularly in rapidly gentrifying neighborhoods.

Entrepreneurship and Small Business Development:

Case Study: The Minority Business Development Agency (MBDA)

The Minority Business Development Agency (MBDA) is a federal agency that promotes the growth and competitiveness of minority-owned businesses through technical assistance, access to capital, and business development services. By supporting minority entrepreneurs and businesses, the MBDA fosters economic empowerment, job creation, and wealth generation in underserved communities.

Education and Workforce Development Programs:

Case Study: Year Up

Year Up is a national nonprofit organization that provides low-income young adults with training, internships, and professional development opportunities to launch successful careers in high-demand industries. Through partnerships with

276

employers and educational institutions, Year Up equips participants with the skills, networks, and credentials needed to access well-paying jobs and advance in the workforce.

Impact Investing and Social Enterprise:

Case Study: The Calvert Foundation

The Calvert Foundation is a nonprofit organization that mobilizes capital to finance social enterprises, community development projects, and impact investments that address pressing social and environmental challenges. By channeling investor dollars into projects that generate positive social and financial returns, the Calvert Foundation catalyzes sustainable development, poverty alleviation, and wealth creation in underserved communities.

Policy Innovations and Legislative Reforms:

Case Study: The Community Reinvestment Act (CRA)

The Community Reinvestment Act (CRA) is a federal law that encourages banks and financial institutions to meet the credit needs of low- and moderate-income communities, including minority neighborhoods and underserved rural areas. Through CRA compliance, banks invest in affordable housing, small business lending, and community development projects, promoting economic inclusion and neighborhood revitalization.

Philanthropic Initiatives and Impact Philanthropy:

Case Study: The Ford Foundation's BUILD Initiative

The Ford Foundation's BUILD Initiative is a grantmaking program that provides flexible, long-term funding and capacity-building support to nonprofit organizations working to advance social and economic justice. By investing in organizational sustainability, leadership development, and strategic planning, the BUILD Initiative strengthens the capacity of grantees to drive systemic change and address the root causes of inequality.

Conclusion: Case studies and success stories illustrate the diversity of approaches, the power of innovation, and the transformative impact of initiatives aimed at addressing the wealth gap and

creating a more equitable economy in the United States. By highlighting successful interventions across different sectors and communities, policymakers, advocates, and practitioners can draw inspiration, learn from best practices, and collaborate to scale effective solutions that promote economic justice, reduce inequality, and create pathways to opportunity for all individuals and families striving for a better future.

Examples of Effective Policies and Programs

Introduction: Effective policies and programs play a crucial role in addressing the wealth gap and fostering a more equitable economy in the United States. By implementing targeted interventions, investing in human capital, and promoting economic inclusion, policymakers can create opportunities for all individuals and communities to thrive. This chapter examines examples of successful policies and programs that have proven effective in reducing inequality, promoting economic mobility, and narrowing the wealth gap.

The Earned Income Tax Credit (EITC):

The Earned Income Tax Credit (EITC) is a federal program that provides refundable tax credits to low- to moderate-income working individuals and families. The EITC supplements wages, lifts millions of families out of poverty each year, and incentivizes workforce participation. Research shows that the EITC has a positive impact on reducing income inequality and improving economic outcomes for low-income households.

Affordable Housing Tax Credits:

Affordable Housing Tax Credits are a key tool for incentivizing the development of affordable housing units for low- and moderate-income households. State and federal tax credits encourage private investment in affordable housing projects, expanding access to safe and affordable housing and reducing homelessness. Affordable housing tax credits contribute to neighborhood revitalization, promote economic stability, and create opportunities for wealth accumulation through homeownership.

Community Development Financial Institutions (CDFIs):

Community Development Financial Institutions (CDFIs) are specialized financial institutions that provide affordable credit, capital, and financial services to underserved communities. CDFIs serve as a lifeline for minority-owned businesses, low-income individuals, and communities lacking access to traditional banking services. By promoting financial inclusion and entrepreneurship, CDFIs contribute to wealth creation, job growth, and economic resilience in disadvantaged areas.

The Affordable Care Act (ACA):

The Affordable Care Act (ACA) expanded access to health insurance coverage, reduced healthcare costs, and improved health outcomes for millions of Americans. By subsidizing insurance premiums, expanding Medicaid eligibility, and prohibiting discrimination based on pre-existing conditions, the ACA helps families avoid medical debt, protect their assets, and maintain financial stability in the face of healthcare expenses.

281

The Supplemental Nutrition Assistance Program (SNAP):

The Supplemental Nutrition Assistance Program (SNAP) provides food assistance to low-income individuals and families, helping them afford nutritious meals and alleviate hunger. SNAP benefits support household budgets, free up resources for other essential expenses, and reduce food insecurity among vulnerable populations. Research demonstrates that SNAP reduces poverty rates and improves health outcomes, particularly for children and seniors.

Career and Technical Education (CTE) Programs:

Career and Technical Education (CTE) programs equip students with the skills, knowledge, and credentials needed to succeed in high-demand industries and secure well-paying jobs. By offering vocational training, apprenticeships, and industry certifications, CTE programs prepare individuals for career pathways in fields such as healthcare, technology, and skilled trades, narrowing the skills gap and increasing workforce participation.

The Children's Health Insurance Program (CHIP):

The Children's Health Insurance Program (CHIP) provides low-cost or free health coverage to children in families that earn too much to qualify for Medicaid but cannot afford private insurance. CHIP ensures that children have access to preventive care, immunizations, and treatment for chronic conditions, promoting healthy development and academic success. By investing in children's health, CHIP helps families avoid medical debt and financial hardship, supporting long-term economic well-being.

The Pell Grant Program:

The Pell Grant Program provides need-based financial aid to low-income undergraduate students to help cover the costs of tuition, fees, and other educational expenses. Pell Grants enable students from disadvantaged backgrounds to pursue postsecondary education, obtain valuable credentials, and access higher-paying career opportunities. By expanding access to higher education, the Pell Grant Program promotes social mobility and economic empowerment for individuals and families.

Conclusion: Effective policies and programs are essential for addressing the wealth gap and creating a more equitable economy in the United States. By implementing targeted interventions, investing in human capital, and promoting economic inclusion, policymakers can empower individuals and communities to build wealth, achieve financial security, and pursue their aspirations for a better future. Through strategic policymaking and innovative program design, policymakers can harness the power of public policy to promote economic justice, reduce inequality, and create opportunities for all individuals and families to thrive in a more equitable society.

Lessons Learned and Best Practices

Introduction: Addressing the wealth gap and creating a more equitable economy in the United States requires a comprehensive understanding of the challenges, opportunities, and strategies that shape economic outcomes and opportunities for individuals and communities. Through the implementation of policies, programs, and initiatives, valuable lessons have been learned, and best practices have emerged to guide future efforts in narrowing the wealth gap and promoting economic equity. This chapter explores key lessons learned and best practices drawn from successful interventions and innovative approaches aimed at fostering economic inclusion and reducing inequality.

Prioritize Equity and Inclusion:

Lesson Learned: Prioritizing equity and inclusion is essential for addressing the root causes of the wealth gap and ensuring that policies and programs benefit all individuals and communities, especially those historically marginalized or underserved.

Best Practice: Incorporate equity considerations into policy design, program implementation, and decision-making processes to address systemic barriers, promote diversity, and advance economic justice for all.

Invest in Human Capital Development:

Lesson Learned: Investing in human capital development, including education, job training, and workforce development, is critical for expanding economic opportunities, increasing earning potential, and fostering upward mobility.

Best Practice: Prioritize investments in quality education, skills development, and lifelong learning opportunities to equip individuals with the knowledge, skills, and credentials needed to succeed in a rapidly changing economy.

Promote Financial Inclusion and Asset Building:

Lesson Learned: Promoting financial inclusion and asset building is key to empowering individuals and families to build wealth, accumulate assets, and achieve financial security over the long term.

Best Practice: Expand access to affordable financial services, products, and resources, including savings accounts, credit-building tools, and asset-building programs, to help individuals build assets, manage debt, and plan for the future.

Foster Entrepreneurship and Small Business Development:

Lesson Learned: Fostering entrepreneurship and small business development is a vital strategy for creating jobs, stimulating economic growth, and promoting wealth creation in communities across the country.

Best Practice: Provide targeted support and resources to minority-owned businesses, women entrepreneurs, and underserved communities to overcome barriers to entry, access capital, and build sustainable businesses that contribute to local economies and generate wealth.

Address Structural Inequities and Systemic Racism:

Lesson Learned: Addressing structural inequities and systemic racism is essential for dismantling barriers to economic opportunity, promoting racial equity, and creating a more just and inclusive society.

Best Practice: Implement policies and initiatives that explicitly address racial disparities, combat discrimination, and promote equitable access to education, employment, housing, and financial

287

services to level the playing field and
ensure that everyone has a fair chance to
succeed.

Build Partnerships and Collaboration:

Lesson Learned: Building
partnerships and collaboration among
government agencies, businesses,
nonprofit organizations, and communities
is critical for leveraging resources, sharing
expertise, and driving collective action to
address complex social and economic
challenges.

Best Practice: Foster cross-sector
collaboration, establish shared goals and
objectives, and engage stakeholders in
collaborative problem-solving processes to
maximize impact, foster innovation, and
sustain long-term change.

Embrace Data-Driven Decision Making:

Lesson Learned: Embracing data-driven
decision-making processes and evidence-
based approaches is essential for
designing effective policies, evaluating
program outcomes, and measuring

288

progress towards reducing the wealth gap and promoting economic equity.

Best Practice: Invest in data collection, research, and evaluation efforts to track key indicators, identify disparities, and inform policy priorities and interventions that target the root causes of economic inequality and promote inclusive economic growth.

Center Community Voice and Participation:

Lesson Learned: Centering community voice and participation in the policy-making process is critical for ensuring that policies and programs reflect the needs, priorities, and experiences of the communities they serve.

Best Practice: Engage community members, stakeholders, and directly affected populations in decision-making processes, program design, and implementation efforts to foster ownership, accountability, and responsiveness to community needs and aspirations.

Conclusion: Lessons learned and best practices offer valuable insights and guidance for policymakers, practitioners, advocates, and stakeholders working to address the wealth gap and create a more equitable economy in the United States. By prioritizing equity and inclusion, investing in human capital development, promoting financial inclusion and asset building, addressing structural inequities, fostering collaboration, embracing data-driven decision making, and centering community voice and participation, we can build a more just, inclusive, and prosperous society where everyone has the opportunity to thrive and succeed.

CHAPTER 6

Conclusion

"Addressing the Wealth Gap: Solutions for Creating a More Equitable Economy in the US" explores the multifaceted challenges of economic inequality and presents a comprehensive array of strategies, policies, and programs aimed at narrowing the wealth gap and fostering economic equity in the United States. Throughout this book, we have delved into the root causes of the wealth gap, examined its far-reaching implications for individuals, communities, and society at large, and highlighted innovative approaches and best practices for promoting economic inclusion and reducing inequality.

The wealth gap is not merely an economic issue but a profound manifestation of systemic inequities, historical injustices, and structural barriers that have perpetuated disparities in opportunity, access, and outcomes for generations. It reflects entrenched patterns of discrimination, unequal access to resources, and unequal distribution of wealth and power that have marginalized

291

communities of color, low-income households, and other vulnerable populations.

Yet, amidst these challenges, we have also witnessed the power of collective action, policy innovation, and community resilience in driving positive change and advancing economic justice. From targeted interventions like the Earned Income Tax Credit and affordable housing initiatives to grassroots movements advocating for racial equity and social justice, there are myriad examples of effective strategies and successful interventions that have made tangible progress in narrowing the wealth gap and creating pathways to opportunity for all.

As we conclude our exploration of this critical issue, it is clear that addressing the wealth gap requires a bold, comprehensive, and inclusive approach that engages stakeholders across sectors, centers the voices of impacted communities, and confronts the systemic injustices that perpetuate inequality. It demands a commitment to equity, justice, and solidarity, grounded in the recognition that a more equitable economy benefits everyone and strengthens the social fabric of our nation.

Moving forward, we must continue to champion policies and initiatives that expand access to education, promote workforce development, support small businesses, and address the root causes of economic inequality. We must advocate for structural reforms that dismantle barriers to opportunity, combat systemic racism, and build an economy that works for all Americans, regardless of race, ethnicity, gender, or socioeconomic status.

In doing so, we can create a future where every individual has the opportunity to fulfill their potential, contribute to their communities, and achieve economic security and prosperity. Together, let us strive to build a more equitable economy where the promise of America is within reach for all who call it home.

Thank you for joining us on this journey to address the wealth gap and create a more just and inclusive society. The work ahead is challenging, but with courage, determination, and collective action, we can build a brighter future for generations to come.

Summary of Key Findings

"Addressing the Wealth Gap: Solutions for Creating a More Equitable Economy in the US" provides a comprehensive examination of the wealth gap and offers a wide-ranging exploration of strategies, policies, and programs aimed at fostering economic equity in the United States. Throughout the book, several key findings emerge:

The Significance of the Wealth Gap:

The wealth gap is a pervasive and deeply entrenched issue in the United States, reflecting historical injustices, systemic inequities, and structural barriers that have perpetuated disparities in wealth, income, and opportunity.

Root Causes of the Wealth Gap:

The wealth gap is driven by a complex interplay of factors, including systemic racism, unequal access to education and employment opportunities, disparities in wealth accumulation, and inequities in access to financial services and assets.

Impact of the Wealth Gap on Society:

The wealth gap undermines economic mobility, exacerbates social inequalities, and erodes social cohesion, leading to disparities in health outcomes, educational attainment, and quality of life across racial, ethnic, and socioeconomic lines.

Policy Solutions and Programmatic Interventions:

Effective policies and programs play a crucial role in addressing the wealth gap and promoting economic equity. Examples include the Earned Income Tax Credit, affordable housing initiatives, community development financial institutions, and career and technical education programs.

Importance of Equity and Inclusion:

Prioritizing equity and inclusion is essential for addressing the root causes of the wealth gap and ensuring that policies and programs benefit all individuals and communities, especially those historically marginalized or underserved.

Collaborative Approaches and Partnerships:

Building partnerships and collaboration among government agencies, businesses, nonprofit organizations, and communities is critical for leveraging resources, sharing expertise, and driving collective action to address complex social and economic challenges.

Data-Driven Decision Making:

Embracing data-driven decision-making processes and evidence-based approaches is essential for designing effective policies, evaluating program outcomes, and measuring progress towards reducing the wealth gap and promoting economic equity.

Community Voice and Participation:

Centering community voice and participation in the policymaking process is critical for ensuring that policies and programs reflect the needs, priorities, and experiences of the communities they serve, fostering ownership, accountability, and responsiveness to community needs and aspirations.

In conclusion, addressing the wealth gap requires a multifaceted and holistic approach that encompasses policy reforms, programmatic interventions, and systemic changes aimed at dismantling barriers to opportunity, promoting racial and economic justice, and creating a more equitable economy for all Americans. Through collective action, advocacy, and a commitment to social and economic equity, we can build a future where every individual has the opportunity to thrive and contribute to a more just and inclusive society.

Future Outlook and Areas for Further Research

"Addressing the Wealth Gap: Solutions for Creating a More Equitable Economy in the US" lays the groundwork for understanding the complexities of economic inequality and offers insights into effective strategies for promoting economic equity. As we look to the future, several key areas emerge for further research and exploration:

Long-Term Impact of Policy Interventions: Research into the long-term impact of policy interventions aimed at addressing the wealth gap is essential. Understanding

297

how policies such as the Earned Income
Tax Credit, affordable housing initiatives,
and small business development programs
impact wealth accumulation, economic
mobility, and intergenerational wealth
transfer can inform future policy design
and implementation.

Intersectionality and Multiple Marginalizations:

Exploring the intersectionality of race,
gender, ethnicity, and other identities in
the context of economic inequality is
critical. Research that examines how
multiple marginalizations intersect and
compound to exacerbate disparities in
wealth, income, and opportunity can inform
targeted interventions and policy
responses that address the needs of
diverse populations.

Innovative Financing Mechanisms and Economic Models:

Research into innovative financing
mechanisms, alternative economic models,
and community-driven approaches to
wealth creation and asset building is
needed. Exploring models such as

community land trusts, cooperative ownership structures, and impact investing can offer insights into scalable solutions that promote economic inclusion and empower underserved communities.

Technological Disruption and the Future of Work:

Understanding the impact of technological disruption, automation, and the gig economy on the wealth gap is essential. Research that examines how changes in labor markets, shifts in employment patterns, and advancements in technology impact income inequality, job quality, and economic security can inform policy responses and workforce development strategies.

Economic and Racial Justice Movements:

Examining the role of grassroots organizing, advocacy movements, and social justice initiatives in addressing the wealth gap is crucial. Research that explores the effectiveness of community-led campaigns, advocacy efforts, and policy mobilization in advancing economic and racial justice goals can provide valuable insights into strategies for

299

building power, mobilizing resources, and driving systemic change.

Global Perspectives and Comparative Analysis:

Comparative analysis of wealth inequality and economic disparities across countries and regions can offer valuable insights into the drivers of inequality and the effectiveness of policy responses. Research that examines global trends, comparative policy frameworks, and international best practices can inform cross-national learning and exchange and inspire innovative solutions to address the wealth gap on a global scale.

Evaluation of Community-Based Initiatives:

Evaluating the impact of community-based initiatives, place-based interventions, and grassroots organizing efforts is essential for understanding what works in promoting economic equity at the local level. Research that assesses the effectiveness of community wealth-building initiatives, cooperative enterprises, and community development strategies can generate evidence-based insights and inform scalable solutions that empower communities to create and sustain wealth.

Interdisciplinary Approaches and Collaborative Research:

Embracing interdisciplinary approaches and fostering collaboration across academic disciplines, research institutions, and community stakeholders is critical for advancing knowledge and generating actionable insights into the wealth gap. Research that integrates insights from economics, sociology, public policy, law, and other fields can provide a holistic understanding of economic inequality and inform comprehensive solutions that address its root causes.

In conclusion, the future of addressing the wealth gap requires sustained research, interdisciplinary collaboration, and a commitment to social and economic justice. By exploring emerging trends, evaluating innovative strategies, and centering the voices of impacted communities, we can advance knowledge, drive policy change, and create a more equitable economy that works for all Americans.